CAMBRIDGE LIBRARY COLLECTION

Books of enduring scholarly value

Earth Sciences

In the nineteenth century, geology emerged as a distinct academic discipline. It pointed the way towards the theory of evolution, as scientists including Gideon Mantell, Adam Sedgwick, Charles Lyell and Roderick Murchison began to use the evidence of minerals, rock formations and fossils to demonstrate that the earth was older by millions of years than the conventional, Bible-based wisdom had supposed. They argued convincingly that the climate, flora and fauna of the distant past could be deduced from geological evidence. Volcanic activity, the formation of mountains, and the action of glaciers and rivers, tides and ocean currents also became better understood. This series includes landmark publications by pioneers of the modern earth sciences, who advanced the scientific understanding of our planet and the processes by which it is constantly re-shaped.

Memoir of Leonard Horner, F.R.S., F.G.S.

Leonard Horner (1785–1864) was a prominent geologist, educator and, later, a factory inspector. In 1833 he was appointed to the Royal Commission on the employment of children in factories, and he inspected sites around the north of England. His earlier scientific work saw him elected a fellow of the Royal Society in 1813, and he was twice president of the Geological Society. The two-volume *Memoir of Leonard Horner*, edited by his daughter, Katharine Lyell, and published in 1890, is a selection of letters to and from his family and friends. The correspondence gives vivid insights into the world of this influential reformer. Volume 1 focuses on Horner's life from his childhood until 1838, taking in many of the key events of his professional career, including his election to the Royal Society, his attempts at improving higher education in Edinburgh and his involvement with the inspection of factories.

Cambridge University Press has long been a pioneer in the reissuing of out-of-print titles from its own backlist, producing digital reprints of books that are still sought after by scholars and students but could not be reprinted economically using traditional technology. The Cambridge Library Collection extends this activity to a wider range of books which are still of importance to researchers and professionals, either for the source material they contain, or as landmarks in the history of their academic discipline.

Drawing from the world-renowned collections in the Cambridge University Library, and guided by the advice of experts in each subject area, Cambridge University Press is using state-of-the-art scanning machines in its own Printing House to capture the content of each book selected for inclusion. The files are processed to give a consistently clear, crisp image, and the books finished to the high quality standard for which the Press is recognised around the world. The latest print-on-demand technology ensures that the books will remain available indefinitely, and that orders for single or multiple copies can quickly be supplied.

The Cambridge Library Collection will bring back to life books of enduring scholarly value (including out-of-copyright works originally issued by other publishers) across a wide range of disciplines in the humanities and social sciences and in science and technology.

Memoir of
Leonard Horner,
F.R.S., F.G.S.

Consisting of Letters
to his Family and from Some of his Friends

VOLUME 1

EDITED BY KATHARINE M. LYELL

CAMBRIDGE UNIVERSITY PRESS

Cambridge, New York, Melbourne, Madrid, Cape Town, Singapore,
São Paolo, Delhi, Dubai, Tokyo, Mexico City

Published in the United States of America by Cambridge University Press, New York

www.cambridge.org
Information on this title: www.cambridge.org/9781108072847

© in this compilation Cambridge University Press 2011

This edition first published 1890
This digitally printed version 2011

ISBN 978-1-108-07284-7 Paperback

LEONARD HORNER.
(From a Crayon Drawing by Samuel Lawrence.)

MEMOIR OF

LEONARD HORNER,

F.R.S., F.G.S.

CONSISTING OF LETTERS TO HIS FAMILY
AND FROM SOME OF HIS FRIENDS.

EDITED BY HIS DAUGHTER

KATHARINE M. LYELL.

IN TWO VOLUMES.—VOL. I.

[PRIVATELY PRINTED.]

London:
WOMEN'S PRINTING SOCIETY, LIMITED,
GREAT COLLEGE STREET, WESTMINSTER, S.W.
1890.

He spent his life in active exertion in doing good to his fellow-creatures and his culture and love of knowledge was only second to his benevolence.

CONTENTS TO VOL. I.

CONTENTS.

CHAPTER XI.

1832—1833.

CHAPTER XII.

1834—1835.

CHAPTER XIII.

1837—1837.

CHAPTER XIV.

1838.

Erratum, page 220, line 9, *after* " Dean of Carlisle "
insert " Bishop of London."

MEMOIRS

OF

LEONARD HORNER.

CHAPTER I.

1785.—1811.

[LEONARD HORNER was born on the 17th of January, 1785, at
George's Square, Edinburgh, son of John Horner, a linen
merchant, by Joanna, daughter of Mr. John Baillie. They
had seven children, of whom Francis, born 1778, was cut off,
after a short and brilliant career, at the age of 38. Leonard,
the subject of this memoir, was their third son. Mr Horner
had been originally a silk merchant in the house of Dundas
and Callander in Edinburgh, and when taken into partner-
ship by Mr. Inglis, (afterwards Sir Patrick Inglis, Bart.) he
found the silk mercery quite a ruined trade. He therefore
began a new line of business in the linen trade, and by his
enterprise, diligence and ability he soon created a very
flourishing business, and supplied the wholesale linen
drapers in London with Scotch linen. This continued till
1809, when the fatal effects of Buonaparte's Continental
Blockade, affected this, as well as other branches of
commerce.

Mr. Horner was a man of much ability and culture; he
was a great reader, and possessed a very remarkable memory

and could repeat whole pages of prose, after once reading,
especially if the language and sentiment had impressed him.
He was an earnest politician, and an ardent Whig.   There is
a good portrait of him by Sir Henry Raeburn.   His wife, of
whom there is also a portrait by Raeburn, was the daughter
of Mr. John Baillie, younger son of the family of Baillie of
Dochfour, in Invernesshire, a beautiful place situated high
on the left bank of Loch Ness.   Mr. Baillie had been brought
up to the Law, and was admitted a member of the Society of
Writers to the Signet in Edinburgh, but he gave it up, and
took to farming in East Lothian, in the parish of Gladsmuir,
of which Dr. Robertson, the eminent historian, was the
minister; and there Joanna Baillie was born in 1754, and
was christened by Dr. Robertson.

Both parents took the greatest care of the education of
their children, and they met with their reward.   By his
sister's account, Leonard as a boy was so lively and gay,
that he was the life of his father's house; always merry,
singing sweetly, and adored by the servants.   He had a very
tender heart, and was devotedly fond of his mother.   With
all his gaiety of temper, he was so reasonable and thoughtful
that if any little matter vexed her, she could open her heart
to him, as much as to his elder brothers.   He was most
generous in spending his money, and doing little kindnesses
to others, whenever he had an opportunity.

He was sent to school at the age of seven, and at nine he
entered the High School of Edinburgh; his elder brothers
Francis and John were pupils of the same school, which at
that time was presided over by its celebrated Rector, Dr. Adams.

His intimate companions were Henry Duncan, son of Lord
Duncan, Charles Napier and James Deans, who afterwards
became Admirals.   Leonard was bent upon going to sea, and
it was with difficulty his father dissuaded him from entering
the Navy, but as Mr. Horner had no Parliamentary or other

interests, Leonard could not have got on in that profession, and he was reasonable enough to understand that his father was right. After he grew up, he often said how glad he was that he had submitted to that wise judgment.

In August, 1799, at the age of 14 and a half, he left the High School, and in November of the same year began to attend lectures at the University of Edinburgh, the Mathematical class of John Playfair, and Dugald Stewart on Moral Philosophy and Political Economy. In 1802 he entered as a student Dr. Hope's class of Chemistry, and began to make experiments at home.

When about 18, he joined the Regiment of Gentlemen Volunteers at Edinburgh, then organized from the fear of a French invasion. At this time he began to study the Huttonian Theory of Geology, and his sisters remembered his taking them to Arthur's Seat (a favourite resort of Dr. Hutton) with his hammer.

His father having established a very profitable business, was desirous that his son Leonard should follow the same calling, and at the age of 19, he made him a partner, giving him a small share.

About this time Mr. Horner contemplated moving his residence to the neighbourhood of London, where his son Francis was already established as a lawyer.]

CORRESPONDENCE.

From his Brother Francis.

London,

6th June, 1803.

MY DEAR LEONARD,—Last time I wrote to you I was so much hurried, that I gave you only a few lines; I fear you have refrained on that account, to favour me with another letter, which I have been looking for day after day. I look forward to the pleasure of having you in town, where I shall consider myself honoured to officiate as your *Cicerone*. I have mentioned to my mother my regret in not having got into chambers, that I might have given you a bed under my roof. But to make the best of it as it is, I should like you to have a lodging in this neighbourhood; it would be enough to have a room for a bed, as you would of course breakfast with Brougham and me, and go with us, (when not otherwise engaged) to the coffee houses where we dine.

I must inform you of a sad accident, which punished my want of generosity towards you; my favourite and faithful thermometer was broken to pieces, in my journey up. I then, not before, regretted that I had not gratified your covetousness; for though you said nothing, you rolled a longing eye towards it. When my father gets to his farm, I hope he will give us an outhouse for a laboratory, where we may make discoveries without frightening my mother. Even then, her towels will not be quite out of danger, and we shall have the old battle renewed, between science and diaper. I cannot tell you with what pleasure I look forward to the execution of my father's plan, the evenings are often very dull in Northumberland Street, and I am sometimes half stupid, half sulky, when I dine alone at a coffee house. When I get you in London, and the rest of the family at no great distance, I shall need nothing to make my happiness more complete, unless I could with a wish bring up along with you, one or two of your Scotch barristers, and not to be extravagant, I would limit myself to *two* of them.* You may ask these two if they will come.

> With love to all, I am, my dear Leonard,
>
> Most truly yours, FRA. HORNER.

* John Murray and Francis Jeffrey, afterwards Lords of Session in Edinburgh.

From the Same.

MY DEAR LEONARD,—I am glad to find you have been
reading Adam Smith, it is quite true, as you remark, that
upon the subject of money he is not quite clear, the true
reason of which is, that in some points he is not quite right,
as you will be satisfied after you have studied a little farther.
What I say now applies entirely to the 5th chapter, which I
would not have you puzzle yourself with too much; it is
sufficient for your present purpose, that you take from it the
general idea of what distinguishes the money price of commo-
dities from their *real* price, or their price in other goods.
This will fully enable you to understand all his subsequent
reasonings, with regard to prices, at least all such of them
as are accurate, and quite free from error, for you are not to
take for granted all that is given you in this book, more than
you ought to do in any other; and it should be your rule in
reading, upon all occasions and all subjects, to examine the
truth of every argument by the force of your own understanding. There is less chance, however, of being led into
false opinions by the "Wealth of Nations" than by almost
any other book on that kind of philosophy.

With respect to the time you are to divide between political
economy and chemistry, you must of course judge for
yourself. To a certain degree, I believe it is most beneficial
in study, to indulge one's taste and predilections, particularly
if they are strong, but in this, to be sure, some discretion too
is necessary. I shall only say, that you will have *less*
opportunity of being taught the elements of political economy
here, than of prosecuting your knowledge of chemistry, which
I presume by this time is a good deal more than elementary.
There are in London more than a dozen courses of lectures

on chemistry, though none certainly so valuable as those at Edinburgh, but there are no lectures whatever on political science. I have no purpose in mentioning all this, but to make you fully aware of the advantages of your present situation; if you were to neglect them, which I am sure you will not, you might feel a regret when it would be unavailing.

With respect to chemistry in general, I would advise you from my own experience, not to throw away many hours upon manual and operative experiments; because your plans of life will not admit of your ever becoming a perfect, or even a tolerable workman. Indeed your views should be a good deal higher than that; you have surely no ambition to discover a new metal or earth; if you had, you have no chance of success, unless you chain yourself for life to a furnace. Do not imagine that I am disposed to damp your ardour for chemical studies; on the contrary, my design is to promote it by urging you to choose the most important and most difficult parts of the science. Do not quit it till you have mastered all the general knowledge that has hitherto been ascertained, of the chemical phenomena of nature; and have fixed in your mind such a store of facts, and principles, and *reasonings*, as will enable you, after you turn your labour to other subjects, to follow the future discoveries and improvements that may be made. Only remember, that an evening for your purpose may be far more profitably employed in labouring to comprehend a general theory or exposition of great natural phenomena, or to detect the fallacies of an inaccurate hypothesis, than in watching the manipulation of a process, which you perfectly understand, but which you cannot qualify yourself perfectly to execute. You have to render yourself a man of liberal information, and of a cultivated understanding, not a dexterous apothecary, I often reflect with pain and regret, that the hours I have wasted in distilling sulphuric acid for future experiments that never were attempted, might have put me in full and steady possession of the doctrine of latent heat, the theory of affinities, that of oxygenation in its various branches, &c., &c.

Write to me soon, and all about your studies, &c., without minding the rotation of other letters from York Place.

Affectionately yours,

FRA. HORNER.

From the Same.

Temple
November 26th, 1803.

You are quite right, my dear Leonard, in giving yourself up to that branch of science to which your taste is decidedly directed. An ounce of genuine ardour and enthusiasm, is worth a whole ton of mere deliberation and plan. In recommending Political Economy to you at present, I was influenced by the single circumstance of your having but this opportunity of hearing Stewart.* Were you to remain in Edinburgh I should advise you to think of nothing but chemistry, till you have mastered all the general speculations and traced the whole outline of the science. Because I am satisfied that one object is quite enough, and that taking them successively is the shortest, as well as the surest way of getting through them all, but circumstances may sometimes, as yours at present, render it upon the whole, more advisable to sacrifice a little, both of time and of present inclination, for the sake of profiting by an opportunity which will not afterwards be in your power. All this, however, I merely throw out as a hint, of which you are yourself to consider the propriety, for the truth is, study is one of those parts of conduct on which one can very seldom give advice of much consequence to one another. Where there is a real inclina-

---

* Professor Dugald Stewart

tion, it should be left to its own freedom, and I am very far
from thinking of leading yours, because I have always
indulged my own, and imagine that I have profited by that
independence.

I can easily pardon you for not being an enthusiast, as
you call it, in political economy; at the first look it is somewhat crabbed and unentertaining. Then it seems probably
to have too much ado with pounds, shillings and pence, of
which you have enough in your hours of business, and
wish to forget them when you come to amusement.
At present I will refrain from attempting to undeceive
you in this respect. The doctrine of heat is indeed very
curious and beautiful, perhaps the most interesting in the
whole range of chemistry, because it is connected with all the
great phenomena of every other branch. As yet, however,
there is a great deal of erroneous reasoning about it, and one
must be very cautious in adopting any of the prevailing
theories. The question as to its materiality is far from being
decided by good evidence either way and perhaps it is a
question which savours too much of metaphysics, and belongs
rather to that science, than to experimental philosophy. I
should think it most useful and most logical to state the
ascertained truths with respect to the phenomena of heat, in
such a form of expression as should leave that controversy
altogether out of view. There is an excellent memoir on the
general doctrine of heat in the Memoirs of the Academy of
Sciences, for 1780, written by Lavoisier, and, I believe,
Laplace in conjunction. The views these have taken seemed
to me some years ago, more precise and scientific than any I
had got elsewhere, but I confess there is some presumption in
my talking upon these subjects; they have been so long out
of my head, and I made so little of them when they occupied
me exclusively. But I will venture to recommend one
practice to you with confidence; that is, writing upon what-

ever subject of investigation engages you, making notes of
whatever conjectures or criticisms occur to yourself, and
before you leave the subject, drawing out a methodical
sketch of the general truths, to which you have made up your
mind as proved, expressing them in your own language,
which will of course be plain, concise and accurate. This
practice, I assure you, saves a great deal both of time
and labour. It sweetens a little the toil (it is always such) of
thinking for one's self; and renders it more easy to read with
great attention, which is a habit so difficult to acquire, that
one must facilitate it by all means, direct and indirect.

There is not a better subject to try this practice upon, than
that of caloric, with which you are at present engaged.

I have been living so much out of the scientific world
lately, that I know nothing of what is going on. But so
much remains of my good Edinburgh habits, that you have
inflamed my curiosity about the expansion of freezing water;
you are morally bound, therefore, to quench this flame. I
request you will tell me what the point is, upon which the
controversy turns, and what are the most important proofs
of experiment that have advanced, *pro* and *con.* But I want
no theories about forms of the crystals, included air, &c., for
I can make a hundred such for myself any night in the dark.

I am glad to hear that your friend, John* Brougham, has
got a settlement at last so much to his mind. I sincerely
wish him success, as I am much interested in the prosperity
of every member of that family, to which I am attached, not
merely by long habits of acquaintance, but by an intimate
knowledge of their great worth. Henry, I hope, we shall
soon have among us here; and if Murray and Jeffrey could
be transplanted also, my comforts would be complete, and
we might renew at Frognal, under my good mother's
hospitality, some of the old parties, with the learning and

* Brother of Henry Brougham

chattering of which you were not always quite satisfied, until oxygen and retorts made a conquest of you to our side.

When I was at Oxford last week, I attended an evening lecture on *Galvanism*, which lasted two hours, by Dr. Kidd, the reader in chemistry, an old acquaintance of Lord Webb's.* I was really much pleased with it. It was exactly what it was intended to be, popular and interesting, adapted to the capacity of young gentlemen commoners, among whom I was pleased to see germinating Lords and embryo Statesmen attending with apparent satisfaction and diligence.

The whole of the present notions of Galvanism were explained in this lecture, and all the leading experiments peformed, both upon animals recently killed, and upon the metals. Some of them I never before had an opportunity of seeing performed. They are all highly interesting, some beautiful from their coincidence with Lavoisier's system, others puzzling by their discrepancy. If I durst return to chemistry, I should attack this puzzle instantly. *Eheu fugaces.* Every man dreams of what he is to do, when he has secured his independent competency. I allow myself to dream of returning to science. But in the long interval that must be toiled through first, I may perhaps be vulgarised enough to forget it, and gather other subjects of dreaming. At the present it is enough for me to wish that in the interval *Nec sit inutilis toga, Nec indiserta lingua nec turpis manus.*

I am, my dear Leonard,

Faithfully and affectionately yours,

FRA. HORNER.

[In the winter of 1803-4 Mr. Horner took a house at Frognal in Hampstead. His business in Edinburgh was carried on by his partner, Mr. Baxter, and that portion which

*Lord Webb Seymour, see Memoir of, in Appendix, to Francis Horner's life.

had always been confined to London was conducted by himself with the assistance of his son Leonard.

In the spring of 1803, before he left Edinburgh, Leonard made the acquaintance of Samuel Tertius Galton, who had been sent by his father to pass a winter at Edinburgh University. This acquaintance gradually warmed into a friendship which continued with their lives.* Francis Horner came often up to his father's house at Hampstead, bringing with him various friends. Among others James Brougham (younger brother of Henry, afterwards Lord Brougham) who had been trained to be a Writer to the Signet in Edinburgh, in the office of Crawford Tait, father of the future Archbishop of Canterbury.

In October, 1804, Leonard made the acquaintance of his future wife, Miss Anne Susan Lloyd, the daughter of Gamaliel Lloyd, Esq., lately come to reside in Hampstead, and whose landed estates were in Yorkshire; she was not quite 18, and they were married in June, 1806, when he was 21.

They lived together nearly fifty-eight years, and there never was a happier marriage. She was very beautiful, and was remarkable for her purity of principle, her liberality of mind, her affectionate heart, and her entire absence of selfishness. Perhaps her most distinct characteristic was her truthfulness, which pervaded the whole of her inner and outward life.

In the spring of 1805, Leonard spent some months in Edinburgh on business matters. Here he took lessons in Italian, and afterwards made a riding tour through the Highlands with William Horton Lloyd, the brother of his future wife, and they were accompanied on their tour by Dr. John Gordon. They rode by slow journeys back to England, stopping at Warwick, where Leonard first made the acquaintance of Dr. Parr, who received him with great favour owing to his Whig connections.

* S. T. Galton was the father of Francis Galton, author of "Hereditary Genius," &c.

From his Brother Francis.

Temple,

June 28th, 1805.

MY DEAR LEONARD,—You have learned from the newspaper what course Lord Melville's affair has taken. The impeachment is unquestionably the true mode of proceeding, but the way in which it has been got at, seems most disgraceful to the House of Commons and calculated very much to lessen its character in the country. All that disgrace however, should fall on the head of Mr. Pitt, whose want of plain conduct, if not want of firmness, in the whole affair of Lord Melville, has been equally prejudicial to Lord Melville's chance of escape, to his own character, and to the constitutional dignity of such proceedings. The Impeachment will scarcely be brought on this session, because preparations must be made in the Hall; though an attempt is said to be intended, that it should take place in the ordinary room of the House of Lords. But this, I presume, will be resisted by the House of Commons; at least it ought to be, for the publicity of such trials is perhaps the best part of their importance. There will be three articles of charge, which will be presented in a day or two, and I hope Lord Henry Petty* will have to conduct one of the three. We hear a great deal of the squabbling among the Ministers: that is, the Doctor's† friends against Pitt's. They can hardly go on very long together, if they are let alone; though their common interest will make them try reconciliation as much as it is practicable. The King is said to be losing his conceit of the Doctor; who is no longer (he says) a good plain man but setting up for a *politician* like the rest of them. This I take to be about the best thing his sacred Majesty has uttered during his happy state of inspiration. Lord Sidmouth again observed to Kinnaird the other day, that there is a great deal

---

* Afterwards Lord Lansdowne.

† Lord Sidmouth.

of phlogiston in the House of Commons, which you as a chemist will understand to be the Doctor's way of describing that restless, uncertain, changeable state of votes which proves very troublesome to quiet Ministers who would be content to go on in their own way.

Yours truly,
F. H.

[In 1808 Mr. L. Horner became a member of the Geological Society of London, and from that period the progress of Geology, and the prosperity of the Society, claimed his most zealous attention, and became the chief occupation of his leisure hours.

In 1810 he was chosen one of their two Secretaries, and used to spend many evenings arranging the collections, frequently accompanied by his young wife.

About this time Mr. John Horner with his family left Hampstead, and took a house in Russell Square, where they were surrounded by their nearest relations. They had also many delightful friends, political, literary and scientific. Sir Samuel and Lady Romilly, and Dr. and Mrs. Marcet, lived in Russell Square, Mr. Courtenay, (afterwards Earl of Devon) and Lady Harriet, and Mr. and Mrs. Hallam lived in Bedford Place. They also saw much of Mr. William Murray and his brother John (afterwards Lord Murray), Mr. and Miss Edgeworth, and Mr. James Abercromby. Besides these they knew many foreigners, most especially Genevese, both at Dr. Marcet's and their own houses, among them M. Dumont, the friend of Jeremy Bentham and the biographer of Mirabeau.

On the 9th of October, 1808, their eldest child was born, and named Mary Elizabeth,* and early the following year Leonard went to Edinburgh and afterwards to Ireland on a business tour.]

* Afterwards the wife of Sir Charles Lyell, Bart.

To his Mother.

Holyhead,
June 21st, 1809.

I AM delighted to set foot again in England, though yet only as far as Wales, but I return delighted with Ireland.

The country is beautiful, and the people the most kind and hospitable in the world. I have been loaded with kindness everywhere, and they would hardly let me off. Whenever I have time and money to travel, I shall go first to Ireland. It is a disgrace to the people of Britain to know so little of it as they do, and the usage which Ireland receives from England is felt by all ranks.

[In September, 1809, Leonard and his wife and brother Francis, made an excursion together.

Francis writes from London, " I have projected a little journey with Leonard and Anne for a week, from which I promise myself much pleasure. We are to go to Birmingham by Warwick, upon a trip, half for political economy and half for geology. The manufactures of Birmingham will give us a taste of the one, and for the other we mean to study the country which Mr. Playfair has described in his book. Warwick Castle you will observe will throw a dish of the picturesque to season the whole."

During the summer months for several years Mr. and Mrs. L. Horner, with their little girl, were in the habit of visiting different parts of England. In 1810 they spent some time at Malvern and went over to Droitwich. His observations on these two places were printed in the Geological Transactions, and excited considerable attention.

At the time, and after Mr. Horner's death, in the Presidential Address at the Geological Society in 1865, Mr.

William Hamilton alludes at length to them, and of their geological value.

After describing another paper written a few years later, Mr. Hamilton goes on to say "It is impossible to read these early papers of Mr. Horner's without admiring the cautious manner in which he avoids a too hasty generalization.

"He clearly saw, as it were glimmering in the distance, some of the great principles and laws which have regulated the order of superposition, and the stratification of the various formations, but the great laws of palæontology, and the now well-known order of succession in the history of organic life, were then only just beginning to be understood. Facts were wanting to enable the geologist to understand the true order of superposition, and we thus find Mr. Horner carefully avoiding theories, but anxiously endeavouring to collect all the facts he could, for the use of those who might follow him, and thus helping to lay the foundation of those principles which have, chiefly by the labours of Murchison and Sedgwick, been so successfully applied to the history of the Palæozoic rocks."

Professor Playfair, in the *Edinburgh Review* of 1811, speaks of Leonard Horner's paper on the Malvern Hills, and concludes thus :—"We must remark of Mr. Horner's paper, that it is more complete in its accompaniments, than any other in this volume, being illustrated both by a map of the country and sections of the rocks."]

To his Mother.

Ladywood,* near Birmingham,

November 2nd, 1810.

MY DEAREST MOTHER,—Since you heard from us, I have had an extremely pleasant trip into Derbyshire for a few days, from which I returned yesterday. Miss Harriet

* The residence of Mr. S. Tertius Galton.

Darwin, our amiable hostess' very agreeable sister, who you
know has been on a visit here for some time, set out on her
return home last Monday. She had her own horse, and was
to be attended as far as Lichfield by Mr. Galton's servant,
but as she and I had become very intimate, and pleasant to
each other, I could not resist the temptation of offering my
services in the place of the servant; these were kindly
accepted, and accompanied with a pressing request to
accompany her home to the Priory, near Derby, which I
agreed to. We accordingly set out together on horseback
last Monday morning, and after a pleasant ride, arrived at
Lichfield at her brother's house, who has lately settled there
as a physician.* Miss Darwin and I set out next morning,
and got to Burton-on-Trent and from thence to see the ruins
of Tutbury Castle where Mary Queen of Scots was confined.
We next pursued our journey to Derby, stayed an hour there,
and reached the Priory about seven o'clock in the evening.
It is an extremely pretty place, about six miles north-east of
Derby, situated in a wooded valley, above three miles from
any great road. I met with a most hospitable reception from
Miss Emma Darwin. . . . . Her hospitality was very
welcome, for it was a very cold evening, and a great deal of
snow had fallen; we had rode 34 miles, and tasted nothing
from breakfast till that time.

After walking about the Priory grounds next morning the
two young ladies and I set out on horseback to Derby, where
they were anxious to show me as many Lions as my time
would permit.

The most pleasing sight I had, was the new Infirmary,
which was only opened last June. A very large subscription,
£25,000, was collected in the county, and they have erected
a most elegant building, and in its interior arrangement there is everything that can contribute to the comfort

* Dr. Robert Darwin, father of Charles Darwin.

of the patients. By the assistance of a Mr. Strutt, a very able and wealthy philosopher, who lives in Derby, the principles of science have been so ingeniously applied in fitting up the different parts of the building that a system of economy has been introduced which could not in any other way have been carried into effect. It is considered quite a model of what a hospital should be. We returned to the Priory to dinner. I spent a very pleasant evening with the ladies, and next morning I mounted my horse at seven o'clock, to return home, which I reached to dinner at four o'clock after rather a weary ride of 46 miles, the tedium of which was only alleviated by the pleasant prospect of returning to my two darlings* and our kind friends at Ladywood.

From Francis Horner.

Torquay,
September 21st, 1811.

. . . . Your notions of the benefit to be derived from the study of Mr. Playfair's great work,† perfectly coincide, I perceive, with my own. It would be a good exercise for you, and not the less so for being a pretty hard task, to go over the whole work with no other purpose than to examine very strictly the accuracy of the reasoning, to detect (if there be any such faults) where the inference is larger than the evidence on which it rests, and where it is not quite so large as the evidence would warrant: you might as you read along, consider also, at every step, what experiments or observations ought to be made to check such an inference, to make up the supplement of such a set of facts, &c.

If you do not apprehend me clearly, I will explain my

---

* His wife and little girl.
† Illustrations of the Huttonian Theory.

meaning more at large. I would try the justness of every
for, therefore, because, it follows, hence, &c., in every page.
After you have tried this a little while so as to have got into
the way of it, I would have you read the " Novum Organum" of
Lord Bacon, and learn the second book of it, as doggedly and
as much by rote, as if it were the grammar of a new language.
If you will do this, I am sure you will derive immense benefit
from it. It is a winter's work, however; at present your
studies of geology should be all in the open air. Give my
kind love to Anne, I will write to her soon, and am happy
you have had so pleasant a tour.

Yours affectionately,

FRA. HORNER.

CHAPTER II.

1812—1813.

From Francis Horner.

Bath,
April 5th, 1812.

MY DEAR LEONARD,—We have at last two fine spring mornings, though yesterday did not end without a good deal of rain. I had a pleasant walk up the face of Mendip, not long before sunset, just out of Wells: but the wet confined me to the chaise for all the other weary hills—I never knew the ground so soaked as it is, even the steep slopes round Exeter, which one would suppose could scarcely hold water, were like so many quagmires.

I begin to be impatient to get back to town, both to see you all, and to be again in the midst of things, for this is likely to be an eventful summer. The activity of the French armies and fleets, portends some great measure that has been long in contemplation, and this is to be met on our part, by the wisdom of Percival and the Doctor, and Lord Castlereagh. It is whimsical enough, that we are brought back by the Prince to the same administration that in 1803 and 1804 was the ridicule of the whole world, with the difference only of Percival changing his place from the law to the finance, and of our having young Dundas instead of St. Vincent to superintend the distribution of the naval forces. At home we are to have this summer an attempt to raise another Protestant cry, and probably a dissolution of Parliament. In that event, I shall most likely take my leave of it for some time, and though this is of course for your confidence only, I have made up my mind, if I am left out at all, to the determination of staying out till I have done something for myself in the law. I would rather indeed go on upon my

present system, of uniting both pursuits, but in case of a
break in one, I am resolved to give myself up for a considerable interval entirely to the other.

Believe me, my dear Leonard, with kindest love to Anne
and Mary.

Yours affectionately,
FRA HORNER.

[In August, 1812, the Leonard Horners went to Scotland,
leaving their little girl with her aunt, Mrs. Winthrop; it was
the first time Mrs. L. Horner had been there, and they spent
some weeks principally at Edinburgh. They visited Mr. and
Mrs. Dugald Stewart at Kinneil, Mr. and Mrs. Fergusson at
Raith in Fifeshire, and Mr. and Mrs. Crawford Tait in
Clackmannanshire.]

From Mr. Playfair.

Edinburgh,
December 12th, 1812.

MY DEAR SIR,—I am very sorry that the *Essai,* &c., has
been so long of making its way to London. It has, however,
I hope reached the Geological Society before this time. The
circumstances you mention of a fresh water formation being
discovered on the chalk of the Isle of Wight, or, as I suppose
above the chalk, is extremely interesting. Do you know if
Sir H. Englefield's book is in any forwardness.

I hope you found all your friends well on your return to
the South and that Mrs. Horner and you reached the end of
your journey in perfect health. Is your brother quite well?
Is he likely to come into Parliament? I hope he is, though
I remember you almost convinced me that it was better he
should not.

We have nothing new here, I have just been told of an
experiment made by a young man of this place, in which alcohol

has been frozen—the degree of cold has not been measured
as the thermometer always burst. The freezing point, however,
I think may be at least nearly ascertained. However I do
not know enough of the fact to speak with any precision.

I am, Dear Sir,

Yours with great esteem,

JOHN PLAYFAIR.

[In the summer of 1813, on account of the decline of the
linen trade, Mr. Horner's father thought it desirable to
return to live in Edinburgh.]

To his Mother.

Enmore,
October 6th, 1813.

MY DEAR MOTHER,—It is a very long time surely since I
have written to you, at least it appears to me a very long
time since I had a gossip with you. How much more
delightful would it be, if we could talk together. As being
the most acceptable thing which I have to communicate,
I will say first that we are all quite well. Your little darling
is in high health and spirits, fresh and blooming. I will
next say that we saw Frank this morning at Taunton,
perfectly well, and we left him in the act of cross-examining
a tanner, upon the value of raw hides. You have heard of
the pleasant visit we paid to the Acland family at Fairfield.
Since that time I have been making some distant excursions
for three or four days with Dr. Winthrop. We went on
Michaelmas day to breakfast at Stowey, and thence proceeded
to the sea coast, where I had a good day's work with my
hammer. We returned after a pleasant day's journey to
Stowey, where we did ample justice to the orthodox dish of
the day, an excellent goose. Next day we had a long and
very beautiful ride through the Quantock hills, nearly as far
as Minehead. We went with Mr. Poole to visit a Mr. Tripp
at Orchard Wyndham, a fine old place, the property of the

Earl of Egremont, and which belonged to the celebrated Sir
William Wyndham. The house is above 400 years old. Mr.
Tripp is a barrister of eminence, who lives the greater part of
the year in London; he is a very sensible man, and of *course*
a great whig, and of *course* a kind hospitable man. Frank
knows him very well. We spent the night at his house, and
returned the next day to Enmore.

Being very desirous of seeing Frank, we determined to go
to Taunton during the Quarter Sessions, and accordingly
yesterday morning Dr. and Mrs. Winthrop, Anne and I set off,
leaving Mary under the care of our kind friends at the
Parsonage. Mr. Acland had arranged matters very comfortably for us before our arrival; so that we soon found
ourselves in good quarters. We went immediately to the
Court, and after a very sensible charge to the Grand Jury by
Mr. Acland, the business commenced, and Frank entered the
Court in his masquerade dress, and saw us ranged before him
on the Magistrates' bench. He had a large bundle of briefs
in his hand, which he soon began to make use of; he is one
of the two leading counsel, and was engaged in every cause
that came on, while we were there. His manner is exceedingly good; better, I thought, than any of his learned brethren.
We spent the greater part of the morning in Court, the rest
we employed in walking about Taunton, which you know to
be a very fine old town, Frank and William Adam dined with
us at our lodgings, and in the evening we all went to the ball.

This morning we again went to the Court, and when we left
it to return home, we left Frank very busily engaged. He is
looking very well indeed. He expects to be able to leave
Taunton on Friday morning, and to dine at Minto on
Tuesday, so that you will see him on Wednesday next, this
day week. We have had a most pressing invitation to spend
some more days at Fairfield, and accordingly on Saturday we
take leave of Enmore to proceed thither. From Fairfield we

shall go to Wells, but on our way, or rather a little out of our way, Anne and I, and our little pet, go to King's Weston, the seat of Mr. Dickinson, one of the Members for the County, where we shall probably be a couple of days. We shall be about ten days at Wells, and then we go to Bowood, where we shall probably be a day or two. I believe we have told you that Lord Lansdowne wrote to me from Scotland, about a fortnight ago, saying that he had heard of our being in Somersetshire, and requesting us to pay them a visit on our return to town. I intend to be in Gower Street certainly by the 3rd or 4th of next month, to set to work again. You see that in place of the quiet retirement of the country, during which so much reading was to be going on, we are living in a continued round of gaiety, but how is it possible to resist such agreeable people, and such agreeable places. We are certainly surrounded by the most hospitable people that can exist; you must never listen to the boasted superiority of Scotland or Ireland over England in point of hospitality,— it is not true. This is a glorious land of plenty;—when I buy land (if ever I should be able to purchase a cow's grass) I think I shall look at the emerald fields of Somerset. The continued good accounts of yours and my father's health, you may believe gives us real pleasure. Take very great care of yourselves, that you may return to us in the spring fresh and blooming. I am grieved indeed to hear that my poor nurse has lost her son. Poor thing, she has had many severe afflictions. I wish I could afford to make her more comfortable. Perhaps as she and I grow older, I may be able to add to her comforts as they become more necessary for her. Give our affectionate love to my father and sisters. l hope the latter are enjoying themselves very much. God bless you, my very dear mother.

Ever most affectionately yours,

LEONARD HORNER.

To his Mother.

Gower Street, London,
November 11th, 1813.

MY DEAR MOTHER,—You know that I never can forget you for a moment, so do not suppose that my not writing to you sooner is any proof that you have been out of my mind. Anne was going to write to you to-day, but I asked her to let me do it in her stead. We did not want proofs of the kind and affectionate feelings of you all towards us, for we have been so long a time in the constant receipt of them, but such unbounded generosity and attachment is very rarely to be met with, and is much beyond what we can ever repay.* How few fathers there are who would have acted as mine has done, on the present occasion. It is not his generous bounty alone I feel, but the kind, the delicate way in which he conveyed it—no reproaches, no harsh words—not even an admonition—so tender has he been to my feelings; I trust I shall never be so wanting in right feeling as to forget for a moment what I owe to so very good a father. And your letters, my dear mother, are like yourself—can I say more of them? They only express what I have been in the daily habit of receiving from your warm benevolent heart; your's and my father's letters have effected what you intended they should, they have bestowed more comfort than anything else could have done, and have lightened the burden of our disappointments much indeed. We are none of us in low spirits, we have too many blessings to allow so great a weakness to get the better of us; but we are naturally anxious to do what is right, and first to know what is right in the change which it has become necessary for us to make. I have the best advisers, and I am quite disposed to listen with great deference to their advice, and whatever is decided

---

* Leonard had become an Under-writer in Lloyd's Insurance Office, and there were some serious losses which caused him great anxiety, feeling he should not have incurred such a risk.

upon I am sure to meet with a sweet, smiling, contented countenance in my dearest wife, a treasure above everything the world can afford.

Our dearest Mary is quite well, she is in uniformly happy spirits, and becomes daily more intelligent and engaging. Her taste for reading continues unabated, and I am equally happy to say that her passion for dolls rather increases than diminishes. Her learning goes on rapidly, at least as much so as she chooses, for you know we have no desire to cram her, nor to make her a prodigy, but we think it would be wrong not to cultivate carefully her natural intelligence. She is much pleased with geography, which I give her a lesson in, every day between dinner and tea. She is daily giving proofs of that delicacy of feeling which you have often known her to show.

Friday—Frank is very well and very busy. I saw him this morning engaged in a cause in the Court of King's Bench. We see as much of him as he can find time for— and to give a greater chance, we have changed our dinner hour to 6, the time he said would be most convenient.

Among the disappointments that have happened, and the great and distressing changes, that I fear must ere long take place, it is very grateful to me to have received an honour which I did not aspire to. I was last night elected a Fellow of the Royal Society, and I was proposed in a way, every way flattering and gratifying to my feelings, by persons whom I have been proud to gain as friends.

Give my affectionate love to my father, and sisters.

Always most affectionately yours,

LEONARD HORNER.

From Mr. Francis Horner to his Sister.

London,
November 10th, 1813.

MY DEAR NANCY,—I have just had the pleasure of reading your letter after my return from the House of Lords, where I have been all the morning. Leonard is now in much better spirits, chiefly owing to the repeated kind letters he has had from my father. His uneasiness was quite out of all proportion to the grounds that existed for any feeling of that description; anxiety he must feel, and so he must as long as he is engaged in business of that sort; but I have really a very confident expectation that he will not be a loser at all.

*   *   *   *   *

Larpent is exchanged, and restored to his office; Lord Wellington shewed in the whole business a very favourable and flattering opinion of Larpent, which is just what I expected.

*   *   *   *   *

We of the opposition are all very much satisfied, indeed highly pleased, with the pacific tone and moderation of the speech from the Throne, and *per contra;* there is much praise bestowed by the other side on Lord Grenville's, which I heard, and thought very able as well as right in all its doctrines. So that there never was such harmony and concord since I knew Parliament. How long it is to last is another story. From some things I have heard, I am inclined to believe that the real system of the Cabinet at present is an inclination to peace, and a determination to use all their influence with the allies in keeping them reasonable; and I have further heard, from a quarter entitled to some credit with respect to foreign politics, that up to this moment, there is a good understanding among all of them, and an agreement as to the propriety of offering fair terms to France. Pray tell Mrs. Stewart all this, if you see her in

Edinburgh, for I made her unhappy by my prophecy of continued war.

My ground of incredulity still would be, that peace can hardly be agreeable to Carlton House, however, they may be satisfied for the present with Hanover, which the Duke of Cambridge is going over immediately to take possession of.

The Bourbons are considered on all hands, as out of the question, though it is not many months since our Regent, in the fulness of his heart, promised to restore them. Monsieur went over upon the strength of this assurance (it is said) to Bernadotte, but met with such a reception in that quarter that he came back without delay, and they have been crest-fallen from that day. It is a great object of curiosity now to see what course Bonaparte takes, there never was a drama, even in fiction, so highly wrought, as what is passing in our own times. I expect he will be stout, and not hear of peace, though if the terms offered be free of all insult to the French people, that may prove a hazardous part. But the tones of his last Bulletins, in which he seems to me to express everything but despair, like a man who cannot feel it, makes me look for that part to be assumed by him. Nor should I be surprised, considering the peculiarities of temper he has shown on former occasions, if with this unconquerable spirit he should by degrees break out in all the violence and phrenzy of a bloody tyrant, and in the end exhaust the submission of the French, as by his abuse of power, he wore out the patience of the Germans.

Madame de Staël has asked me to dine with her to-day. I do not know who are to be there, except the Hollands.

Francis Horner to his Father.

Lincoln's Inn,
November 15*th,* 1813.

MY DEAR SIR,—I have now had repeated conversations
with Leonard, and he has had much very anxious consideration in his own mind, upon the question whether he ought to
continue any longer to underwrite as an insurer at Lloyds.
My endeavour has been to prevent him as long as I could
from coming to any definite conclusion upon the point,
because I saw, if he determined upon a change of business,
that another very serious question was behind, what business
was to be substituted for the one relinquished. The result,
however, appears now a fixed resolution to give up Lloyds, as
soon as his present engagements there are discharged. What
has chiefly governed him in coming to the determination is a
sentiment (I think) which he feels strongly, that without more
capital than he has to protect him, it is not consistent with
strict integrity to undertake such an amount of risks, as
makes it worth while to be in the business at all. This is an
argument in which he probably leans more to the side of
caution than is required by the ordinary and established
notions of the mercantile world, but which, though I could
have justified a greater degree of adventure, I felt averse to
combat, because he was erring, (if at all) on the safer side,
and on the side of honour, and because a man of Leonard's
intelligence and good feeling acts much more safely upon
such turns of conduct, by following his own sentiments, and
impressions, than by acting up to advice that does not
concur with them. Besides this consideration, he is
influenced a good deal by another, which, in my way of seeing
the matter, is enough for the decision which he has come to ;

and that is, the perpetual anxiety of mind with which
this business distressed him, an effort which no doubt
depends a little upon the relative strength of one's nerves but
which, once ascertained to be incurable, forms a conclusive
objection to the prosecution of any occupation or pursuit; as
no success in life is worth purchasing at the price of daily
uncertainty and apprehension. The insurance business being
thus given up, it becomes necessary that Leonard should,
without delay, fix upon some other plan for the profitable
employment of his time. It would be greatly for his
advantage in some respects, as well as much to his taste and
inclinations, if such could be devised for him in London.
But upon a full consideration of this in every way in which
we have been able to view it, that does not appear to be
practicable, compatibly with his continuing to belong to
your house, which he of course does not bring into question
at all. He very prudently regards his share in your
business as better than any chance or contingency. And
though it will cost Leonard a severe pang to quit connections
and scenes which are on many accounts so agreeable and so
dear to him, and to surrender for the present, advantages in
point of society which he has been gaining every year, as his
character became more known, he has firmness enough (in
which Anne is not at all inferior to him) to take this step at
once, if these considerations render it proper and necessary.

I have thrown aside, from my views of this most important
question for Leonard, a great many considerations which I
felt strongly at every part, in which my own enjoyment of
life is much interested, and I have tried also not to be
influenced by any of those regards pointing different ways, by
which the present gratification of Anne on the one hand,
and that of all those now in Edinburgh on the other hand,
might be more immediately consulted. It is high time that
Leonard should at length be placed in a situation, in which

he can seriously and systematically employ his active mind for the improvement of his fortune, and that he may look forward at an early period of life, when he shall have done everything in that way that is desirable.

I had the happiness to receive my dear mother's letter to-day, and will write to her soon. My kind love to her and my sisters.

Ever my Dear Sir,

Most affectionately yours,

FRA. HORNER.

[Early in December 1813, Mr. L. Horner went to Edinburgh to ascertain whether it would be prudent for him to'remove his residence from London to Edinburgh where his father, mother and sisters were already settled. During his stay there, he decided to go to Holland on business.

Francis Horner writes to his Father.

November 22nd, 1813.

" On the 21st November Baron Peponcher, and one of the brothers of M. Fagel the Greffier arrived in London, as Deputies sent by the Prince of Orange as Stadtholder. Five of the provinces had declared for him. The French army was suffered to retire without molestation, and took refuge in Antwerp and Maestricht. The Prince was expected to leave London this morning the 22nd of November. It is expected that 4000 troops will be embarked immediately to be followed by more."

Thus Holland was opened to England, and it was on that account Mr. L. Horner wished to visit it, to try and extend the commercial relations of his house to that country.]

From Francis Horner to his Father.

Lincoln's Inn,
December 14th, 1813.

My. Dear Sir,—In case Leonard should have left. you before this letter reaches Edinburgh, I address it to you.

I have only to give an account of what passed yesterday with Mr. Baring.* I have settled with him that as soon as Leonard returns to London, he shall call upon him, and that at that interview Leonard would himself explain his views and wishes. Mr. Baring said that Leonard would find Holland at present very dead, such has been the destruction of capital, and that trade can only be expected to rise there very gradually; in the ports upon the Baltic, he said it is otherwise. When Leonard comes to town, I will call with him at Mr. Baring's and introduce him. I understood that the whole interest and concern of the house of Hope at Amsterdam, are now transferred to Baring.

The late news from Holland of the usurpation of sovereign authority by the Prince of Orange, and the abolition of the ancient constitution of that state, throw a gloom over the prospect that we lately thought so promising. Resistance to the French no doubt, and emancipation from a foreign yoke, is the first object and duty of the Dutch, and of all the other oppressed states, but nothing can be more unjust than to take advantage of this opportunity to subvert the constitutional freedom of an ancient government, and by an act of usurpation, under the stale pretence of an expression of the popular wishes to change a republic into a monarchy of undefined powers and prerogatives. It may in the end be found impolitic too; it is at all events unprincipled, and a bad omen of the disposition with which this country begins to

* Afterwards created Baron Northbrook.

exercise its influence in the pretended work of restoring Europe to its former institutions.

We see few here but partizans of the House of Orange, who all approve of this measure. Baring, I am sorry to say among the rest.

With kind love to my mother and sisters, I am my dear Sir,

Affectionately yours,

FRA. HORNER.

CHAPTER III.

1814.

[Early in this year Mr. Horner paid a visit to Holland connected with his business, and he wrote a journal to his wife, extracts of which are given below. He embarked at Harwich on the 30th January, and after a rough passage reached Helvoetsluis on the 30th February, and the next day proceeded to the Briel.]

JOURNAL.

This is a very neat town with a large church; it is fortified, but not strongly. The principal street, which is at least half a mile long, was ornamented with boughs, and triumphal arches of evergreens, intermixed with artificial oranges, and pieces of orange paper, cut in various shapes; these were erected in honour of the young Prince of Orange, who passed this way from England, and I found the same decorations in all the other places through which he travelled. At the Briel we had to cross the Maas, and I found this ferry the most difficult part of our journey, for the river, though not a quarter of a mile broad at this place, was covered with large masses of floating ice. There was only one small boat, which had to be guided through the ice by the ferrymen, with long poles shod with iron, and after waiting for about an hour until another cargo of passengers who were before us were ferried over, we were crammed into this boat with about twenty recruits for the Militia, who were on their way to the Hague. From the Briel we had to pass over to the island of Rozenburg on the Maas, which is cultivated but very thinly inhabited, there being only a few

c

farm houses upon it; here I first noticed a farmer's wife
dressed in the true costume of the country. At Maaslandsluis
we had to hire a carriage waggon to transport ourselves and
luggage to Rotterdam, a distance of about thirteen miles. It
is a very neat small town inhabited by fishermen who are
engaged in the cod and herring fisheries, and it is the
principal place in Holland for the curing of herrings.
Many of the inhabitants must be wealthy, for I noticed
several very handsome houses, and a great many good shops.
We left Maaslandsluis about half past five in the afternoon.
The road is along the top of a very high dyke, and as there
are no side railings it must be very dangerous at night. In
the fields on the side of the river, there lay immense blocks of
ice, which had been floated over, scattered like boulders
of granite. We passed through Vlaardingen, Schiedam,
famous for the manufacture of gin, and Delfshaven, but as
the sun had set before we reached Vlaardingen, I can say
little of them. We reached Rotterdam about 8 o'clock in the
evening, and we drove to the sign of the "Zwynshoofd"
(Boar's head) in the great market place, but I found the
Royal Hotel on the Quay called the Boompjes, a much better
house. At Schiedam there are more windmills I believe, than
in a whole county in England. They are very high and
consist of a circular tower of brick work, and the sails, which
are very large, do not come nearer the ground than twelve or
fifteen feet.

The first thing I did after my arrival at Rotterdam, was to
go to a bookseller's shop to procure a plan of the town, and a
guide. After breakfast I went to deliver my letters of
recommendation to Mr. Samuel Labouchère,* but I was so
unfortunate as to find him out of town, and not expected for
a few days. I have hired a servant, with whom I have got
very good recommendations; he speaks French, Dutch and

* Mr. Labouchère's brother married a daughter of Sir Francis Baring,
and was the father of Lord Taunton.

German, so that he will be able to continue with me when I go to Bremen, and I have got very comfortable lodgings.

The Hague, *February* 13*th*, 1814.

Last week I was at Rotterdam, and although principally engaged in the object of my journey, I have had opportunities of becoming a little acquainted with that town. The population by the last Census in 1811, amounted to about 54,000 souls. The principal street, called the Boompjes, (little trees) is built along the right bank of the Maas, and is about three-quarters of a mile in length; the Maas at this place being about twice as broad as the Thames at Westminster. There are no houses on the left bank, except a solitary one in the midst of some willows which is used as a house of correction I believe. Parallel with the Boompjes there is a broad street with a canal in the middle of it, which, I should suppose, is not less than 150 feet wide, with a quay on each side at least 50 feet broad. All the principal streets in the town are of the same description, and as they are generally full of ships, and are crossed by innumerable drawbridges, it is very difficult to distinguish one haven from another, for so they are called. The water which fills these canals enters from the Maas at each end of the Boompjes and there the ships also enter. The town is adapted for trade in a most peculiar manner, and indeed everything in it appears to be subservient to that object. The ships lie within a few yards of the warehouses, which are under the same roof as the merchant's house. With a very few exceptions all the lower part of the best houses, *i.e.*, the ground floor, is occupied by the counting house and warehouses, and by a separate entrance you go to the dwelling house above. I was very much struck with the exterior splendour of the houses, for those occupied by the chief

merchants excel in that respect many of the best houses in
the squares of London. The French windows, with very
large squares of glass, have also an air of great magnificence.
The greater part of the houses in Rotterdam are very lofty,
being often five and six stories high, and there is no sunk
floor, as the whole town is built on piles. There is, however, a very indiscriminate mixture, for very frequently a
house like a palace is flanked on both sides by a cobbler's
stall, or an ale house. There is a peculiarity in the mode of
building which has a very odd effect, for the houses, instead
of standing quite upright, incline considerably forward.
This is principally the case with the older houses, those of a
more recent date standing according to the plumb-line.
Where an old and a new house are together, the old one
may sometimes be seen inclining forward fully two feet.
The cause of this odd practice was, I am told, with the view
of preventing the rain from soaking into the bricks and
mortar to the injury of the building. All the houses are
built of brick of a dull red colour, with the door post and
lintels of the windows generally of stone, which is usually
painted of a fawn colour. Chimney pots are not used, but
where it is necessary a wooden apparatus supplies their
place.

On the outside of almost every window in Rotterdam there
is a small looking glass for the purpose of seeing while they
sit in their rooms, what people are passing along the street,
and sometimes by a combination of glasses, they can see from
the upper floors those who ring at their doors (for there are
no knockers) and they can immediately determine whether
the person is to be admitted or not, and this they can do
without being seen themselves. I must not omit the statue
of Erasmus in the market place, who was born at Rotterdam.
It is colossal, and of bronze. From the top of the great
Church there is a commanding view of the surrounding

country, but from the extreme flatness the visible horizon is
not at a very great distance. The uniform flatness of the
country, the number of canals and windmills, the river and
Rotterdam, all produce a striking effect, and a view totally
different from what we ever see in England. The fields
having been overflowed before the frost, there is a large
extent of country round the town which is one sheet of ice.
I have not yet seen a great deal of skating, as the ice of the
Maas when I first arrived was very uneven, but as there has
been a thaw, and as a strong frost has again set in, the
surface has become more smooth, and the skaters will
probably appear in greater force.

There is a favourite amusement here among the young
men, which is very interesting to look at from the singularity
of it. It is sailing in a boat with sails upon the ice. The
boat is placed upon a frame which is shod with iron, and it
is guided by a helm with a very strong sharp iron pike
affixed to it, which is struck into the ice when the boat is to
be stopped or turned. The velocity with which they move is
prodigious, for I was told that with a fair wind, and very
smooth ice, these boats have been known to go from Dordt to
Rotterdam in seven minutes, a distance of thirteen miles,
which is twice as swift as the fleetest race-horse I believe.
So rapid a motion must be very dangerous, not only from the
fear of any obstacle, but from the mere circumstance of going
through the air so quickly at so low a temperature.

I left Rotterdam yesterday morning, to pay a visit of a day
or two to this place. I travelled with my servant in a
Fourgon. The road is most excellent and we got to the
Hague in two hours. The principal roads are paved with
small bricks set on edge, and these are soon covered with a
sufficient quantity of sand to remove the unpleasant feeling
of going on pavement. I found Mr. Labouchère at the Inn,
where he had procured a room for me. The first thing I did

was to deliver the letters of introduction which Lady Holland
had given me, to our Ambassador, Lord Clancarty, and to
Mr. Hoppner, one of the Secretaries of the Embassy.   I saw
Mr. H. first, who was very civil in offering to be of any
assistance to me in his power during my stay at the Hague.
I then saw Lord Clancarty, who having read the letters I
brought for him in another room, bolted suddenly into that
in which I was sitting, said he was very glad to see me,
asked me to dine with him to-day, and disappeared as quickly
as he entered.   In the course of the morning Mr. Labouchère
introduced me to Mr. Van der Hoop, one of the partners of the
house of Hoop and Co. of Amsterdam, who conducted me to
Mr. Falck, for whom I had a letter of introduction.   Mr.
Falck is at present the Chief Secretary of the Government,
and has been one of the most active persons in the late
revolution.   I found him very busy in his bureau, as indeed
all the members of the Government are at this moment, for
they have not only to organize the new Military Levies, but
they have to prepare the constitution by which the government is in future to be conducted.   For this purpose fifteen
persons have been elected from the different provinces of
those who were known to be most competent for such a
task.   When they have completed their work, it is to be
submitted to an assembly of the notables which is to be held
at Amsterdam about the middle of March.   Each Province
is to send a list of their notables to the Prince, who is to
select 600 from among them ; this selection is to be submitted
to the people throughout Holland, who are to be invited to
testify, by their signature, their approbation of the persons so
chosen.   In such an assembly it is not intended that there
should be any discussion, but each part will be put to the
vote separately.   What the constitution is to be, has not yet
transpired, but it will chiefly relate to the fixing of the
powers of the Legislative and Executive parts of the

government in their most prominent functions; it is not meant to enter into any of those matters of detail which will more probably belong to the determination of the legislative body, which the new constitution is to establish. At present the French *code* of laws, and nearly the whole system of government adopted by the French, subsist in force, except those parts of it which were most obnoxious, such as the *Haute Police*, a horrid scourge, a tyranny above all law; and some of the most oppressive imposts, together with the abolition of the monopoly of Tobacco by the Government.

There is not, I believe, as yet any party which has shown itself in opposition to the Prince of Orange. The Dutch have been so glad to get rid of the French, that they do not seem to be considering whether new fetters are not preparing for them, at least they seem disposed to wait until they see the promised constitution.—I understand, with very few exceptions, all the persons are continued who were employed by the French in the several departments of Government, such as Judges, Magistrates, Collectors of the Revenue, &c. This has created some discontent among the partizans of the House of Orange, but it is a proof, I think, of great good sense on the part of the Prince and his Ministers.

This morning I walked to Scheveningen, a village on the sea shore about two miles from the Hague. The road is the whole way through a very fine avenue of trees, in a perfectly straight line, with the village church at the end of it. There are eight or ten rows of trees, chiefly elms, oaks, and beeches, but none of them of great size. On one side there is a house and gardens now in the possession of the Bentinck family, but which formerly belonged to Cats, the most celebrated of the Dutch Poets. Scheveningen is wholly occupied by fishermen, but it does not on that account yield to other Dutch villages in respect of cleanliness. There were some small shops containing toys made by the people of the village, most

of them were too bulky for conveyance, but I have bought a
dog for Mary made of sea shells, which will amuse her as the
production of Scheveningen. The boats made use of by the
fishermen are very large and fitted for all kinds of weather,
and as they have flat bottoms, they can be run ashore without
danger.

The whole shore in this part of the coast is bounded by
sandhills or Dunes, which extend about a mile into the
interior of the country, rising in some places to the height of
sixty or eighty feet above high water mark. They are
covered with a long dry grass, which is called Bent.

I have seen some of the principal churches in the Hague,
it being Sunday. The Great Church, or Groote Kerk, is a
vast brick building and very ugly both within and without ;
the only thing worth seeing is the tomb of Admiral Opdam.*
There was an amazing crowd of people, who were singing the
psalms loud enough to drown the sound of a very large organ.
The Prince was there with his wife, mother, and sisters.
They were in a pew in the church, a little elevated above the
rest, with a canopy of green cloth, but without any appearance of pomp. It was only nine o'clock in the morning, and
intensely cold, and the church like an ice-house, so that I
think their Highnesses had infinite merit in submitting to
such starvation. I heard of a poor unfortunate maid of
honour, who had been dancing until four o'clock in the
morning, lamenting most bitterly to one of Lord Clancarty's
family that she had to get out of bed in these cold mornings
at so early an hour, to attend the Royal Family to church.
The French Protestant Church is handsome, and fitted up
with pews in a more comfortable manner. The Hereditary
Prince was there.

* Distinguished in the war between the English and the Dutch in
Charles II.'s reign.

Rotterdam, *February 20th*, 1814.

I dined with Lord Clancarty last Sunday at the Hague; we had a pleasant party, entirely English, consisting of the family, which includes the gentlemen attached to the Embassy, and two or three strangers. I expected the visit would be stiff and formal, but it turned out quite otherwise. Lord Clancarty seems a good-natured man, Lady Clancarty is very pleasing and of gentle manners. Lady Castlereagh is on a visit to this family, while her husband is at the head quarters of the Allied Armies; she was very good-humoured, and seems pleased with everything she sees. There were also Lord Valetort and his sister, Lady Emma Mount-Edgcumbe. In the evening there came the Countess Hogendorp with her three daughters, and Lady Christina Ginkel, the sister of Lord Athlone. The latter has a pretty voice, which she delighted me with, by singing with a Mr. Chad (the son of a Norfolk Baronet), the well-known air of Malbrouk arranged as a duet.

I employed the following morning in seeing the town, and the celebrated House in the Wood, a palace of the Prince of Orange. It is, certainly, with great justice that the Hague has been highly praised, for I have seldom seen anything more to be admired, and I see it just now under unfavourable circumstances, from the season of the year. The number of magnificent houses is quite astonishing, and certainly there is nothing in London that can be compared with the two streets called the Voorhout and the Vyverberg,

I did not fail to see the prison from which the De Witts were taken when they were murdered, it is called the *Gevangen Poort*, but it does not seem to be looked at with much interest by the inhabitants, for several were rather surprised at my desiring to see it. It is a curious circumstance that the first house in which the Prince of Orange

slept after his landing, was that in which John De Witt lived.
Is this to be considered as an omen of the reconciliation of all
parties? It is a very good, comfortable sort of dwelling
house, and has nothing in it that is worth seeing except the
grand saloon, or Salle d'Orange, which is certainly a very
magnificent thing. In the evening a ball was given by the
principal people of the Hague to the English, and Mr. Falck
was so kind as to take me with him. There were about one
hundred and eighty people present. The Prince and the
ladies of his family seldom go out, but the Hereditary Prince
was there, who seems a good-natured, unassuming young
man; he was dressed in a Hussar's uniform and looked well,
he danced the whole evening, and changed partners each
time, so that he distributed his favours among the young
ladies very equally. There was a very fair proportion of
beauty, but the present dress of the ladies did not appear to
me very becoming, most probably from not being used to it,
rather than from anything in the dress itself. The hair is
all drawn up to the top of the head, and in addition to the
bunch produced by this accumulation of curls, there is an
immense bouquet of flowers. They were almost all dressed
in white muslin. There is a great modesty in their manner
of dressing, and there was not a woman rouged in the room.
They dance well, particularly the waltzes, which are danced
quicker than I ever saw them before, and when the rapidity
of the movement is increased, the dance is called a *Sauteuse*,
and is exceedingly pretty. We had supper about one o'clock,
during which the band played the national airs, " Wilhelmus
Van Nassauwen," and " Al is ons Prinsje nog zoo klein,"
alternately with " God Save the King."

The morning after the ball I went to Mr. Labouchère's
house at Ryswyk, with whom I was to return to Rotterdam.
I found him occupied with a farmer's family, who had had
the misfortune of having their cow-house burnt down during

the night, with nineteen cows in it. It was the whole
property of a little farmer who supplied the Hague with
milk, and there is a strong suspicion that the cow-house was
set fire to by a person who had been turned out of the farm.
Mr. Labouchère has set on foot a subscription for their
relief, and I thought it a good occasion for the English to
show their characteristic benevolence, so I wrote to Mr.
Hoppner to use his influence in aid of the subscription, and
I afterwards learnt Mr. Hoppner collected a good sum.

At Ryswyk I went to see the obelisk, erected in com-
memoration of the peace which was signed at that palace in
1697, where the Prince of Orange had a palace, but which
was pulled down about twenty years ago. I had a very
pleasant drive to Rotterdam with Mr. and Mrs. Labouchère.
Mr. L. is the most gentleman-like Dutchman, and the most
intelligent merchant I have yet seen, and his wife is one of
the most charming ladies I have had the good fortune to
meet with in Holland. I dined last Wednesday with Mr.
Ivermondt, one of the most wealthy merchants of Rotterdam.
He was Mayor when Buonaparte visited it, and went with
the Mayor of Amsterdam to Paris at the baptism of the
King of Rome. He had been in favour with the Emperor,
for he received a very beautiful snuff-box set with diamonds,
with Buonaparte's picture on the top, and he afterwards
received a print of the Emperor, from a painting by David.

The next morning I set out early for Dordrecht, in
company with Mr. Jutting, the brother of Mrs. Labouchère.

Dordrecht, or Dordt as it is commonly called, is a place of
considerable trade, chiefly in flax and in wood, vast rafts of
which come down the Rhine. There were also established
here, some time ago, very considerable manufactures of
sugar from the beet-root, one of the remarkable instances in
which Buonaparte's success in his exclusion of our commerce
was seen. I saw one of these establishments which the

sudden renewal of trade had brought to ruin, for the
commonest sugar they made could not be offered for less
than three shillings and fourpence a pound. It is not nearly
so sweet as that from the West Indies. Dordt contained by
the last census 19,500 souls; it is the oldest town in the
province of Holland, and the birth-place of John De Witt.
At Dordt there is the most handsome church I have yet seen.
It is a cathedral, and the oaken stalls in the choir still
remain finely carved. There is a most beautiful screen of
brass-work and marble, between the choir and the nave,
which last is occupied as the place of worship. There is a
screen of the same sort, perhaps finer, in the great church
at Rotterdam. The pulpit is wholly composed of white
marble, richly carved, with steps to it of the same material,
and the sounding-board is a very fine specimen of carving
in wood. The communion plate and the dish for baptism
are of pure gold, and are worth above five thousand pounds.
It was my intention to have left Dordt the following day,
having completed my business, but Mr. Van Poelien having
kindly pressed me to dine with him, I remained till yesterday
morning. He is one of the most wealthy men in Dordt, and
lives in very good style.

On returning to Rotterdam we met the young Prince going
to the head-quarters of the British army, near Breda. He
was travelling in one of the common post-carriages of the
country.

It is melancholy indeed to see the state of poverty to
which this unfortunate country has been reduced. The
great cause has been the total suspension of trade, by which
means the merchants, having no income, were obliged to live
on their capital, and of course those who with small capital,
but a good credit, had from their business derived a good
income, were ruined; then the reduction in the expenses
of the rich took away trade from the shopkeepers, and the

suspension of their business took away bread from all the lower orders who were in their employment. But the misery was not confined to the trading part of the country, for when Holland was annexed, in 1810, to the French Empire, the first thing Buonaparte did, was to reduce the public debt to one-third, by which means all those who had lodged their money in the public funds were at once, without any previous preparation for so sad a calamity, deprived of two-thirds of their property. This measure was not merely confined to the old debt, but was extended to a loan of eighty millions of florins, which his brother, when King of Holland, had made three years before. This blow fell upon all those who had retired; upon many merchants who, having no trade, had invested their money in the funds, and upon all the charitable institutions of the country, who were obliged to lodge their money in that way. The principal stock is two and a half per cent. In prosperous times, about forty years ago, that stock was as high as a hundred and ten. I saw it quoted a few days ago at twelve and a half, so that a person who at that time held stock to the value of one hundred and ten thousand pounds, is now not worth more than twelve thousand five hundred pounds. In addition to all this, they had to pay heavy taxes, and the better families had constantly to pay large sums to liberate their sons from the conscription. The French besides, carried off their horses and their cattle for the army, so that there is hardly a good horse to be got now in all Holland.

The effects of this dreadful loss of property are, of course, to be seen everywhere. Those who formerly kept a carriage and a handsome establishment, can now hardly afford to keep one maid-servant. The greater number of the most wealthy have put down their carriages, and of five thousand horses of pleasure, which paid the duty at Amsterdam four

years ago, there were not more than fifteen hundred to pay
it last year. The country houses have been sold, and many
of them pulled down for the sake of the materials, and to
save the taxes upon them. For the same reason whole
streets in the Hague have been pulled down, all entertain-
ments and fêtes in private were given up, and people lived
retired in their own families. The number of houses that
were to sell and let was prodigious, and consequently their
value was reduced to a mere trifle. Although in Rotterdam
and the Hague there has been a great change in that
respect, still almost every third house has a ticket upon
it to sell or to let. To give you an idea of the fall in
value, I can tell you what the rise has been within these
two months. A house which Mr. Labouchère could have
had for one thousand four hundred florins a year, the
proprietor now requires two thousand five hundred for,
and the difference is still greater at the Hague, where
the houses have tripled in value, and as the Court is to
be established there again, the demand is expected to be
very considerable. This impoverished condition has, however, not extended its effects very greatly among the country
people, and they all appear well clothed and fed, and their
houses are well furnished. They have had to bear very
heavy taxes, but they have always had a ready sale for
their produce, and the price of labour has not fallen; they
have always had a considerable sum of money circulating
among them for substitutes in the conscription, a father
frequently receiving five hundred pounds for his son's services.
But in the towns the number of poor has increased very
much indeed; in Rotterdam there are at this time eleven
thousand persons who are wholly supported by charity,
which is one-fifth of the whole population. There is no
direct poor rate, but the money is collected in a variety
of forms, there are various duties paid to the town and

from the town's chest a considerable proportion goes to
the support of the poor ; several articles pay a certain duty to
the poor, each box-ticket at the theatre pays four shillings.
At church during a psalm, certain persons of the congregation go round to receive a contribution, and in every
merchant's counting-house I have seen a poor-box.

I left Rotterdam on Thursday, February 24th. I spent
about an hour and a half at Delft ; the town is very pretty,
the streets are spacious, with a canal in the middle, and trees
on each side, which is the case, I believe, with most of the
towns in Holland. There were in it by the last Census
14,000 souls. There was formerly here a very extensive
manufactory of earthenware, but which has long since gone
to decay, the superiority of the English pottery having beat
it out of the market, nor has it increased much again during
the suspension of intercourse with England. The Mausoleum
of William, first Prince of Orange, in the new church, is
certainly a very fine thing, but I was much better pleased
with the plainer monument of Grotius in the same church ;
he was a native of this town. In the old church, there are
the monuments of the Admirals Tromp and Hein. Finding
the door of the Stad-haus open, I went in, and found the
door of the Council Chamber open, where there is a collection
of the portraits of the Princes of Orange, and a very fine
portrait of Grotius, but by what Master I could not learn.

Between Delft and the Hague, we paid a short visit to
Mrs. Labouchère at Ryswyk. I say we, for I am accompanied by her brother, Mr. Jutting, who is to be my
companion during the remainder of my journey through
Holland ; without his assistance or that of some other
interpreter, I should be badly off ; so little French is
understood, but I am studying the language vigorously.

At the Hague we dined at a Table d'Hôte, and during
dinner I was much amused by a boy and girl in the street

singing in parts, and remarkably well; their ballads were
laudative of the Prince of Orange, and abusive of Buonaparte
in the extreme, the airs were very pretty. In the Fish
Market I noticed two storks walking about in full security,
they are, you know, sacred birds here. As it was found that
they returned even when this country ceased to be a republic,
I suppose the veneration for them is rather on the decline.
We reached the famous city of Leyden a little after sunset,
it was intensely cold.

The Burg or Castle, which is situated on an artificial
mound in the centre of the town, sufficiently high for the
cannon to play over the tops of all the houses, is celebrated
in the memorable siege of this town by the Spaniards in
1574; it is now, however, in a state of ruin. The anatomical
Theatre is the only thing worth seeing in the University,
where the collection of the famous Aldinus is kept entire in a
room appropriated for the purpose. The number of students
that attend the University is about 300, and almost entirely
natives of Holland. Mr. Sandifort, the Professor of
Anatomy, and Mr. Siegenbeek, whose attention has been
chiefly directed to the improvement of the language of his
country, are, I believe, the most eminent men of the place.
I did not fail to visit the monument of Boerhaave, in the great
church, but everything in Leyden bears the marks of decay,
its population, which in its most flourishing state amounted
to forty thousand, is now under twenty-two thousand.

There used to be a very extensive manufacture of woollen
cloth, and by the books of the Cloth Hall, thirty two thousand
pieces have been made in a year, and now there are scarcely
one thousand made. The superfine black and blue cloth is
still made of a very superior quality, and is rather dearer
than in England. There are several streets in the outskirts
of the Town, which within these few years have been pulled
down for the sake of the materials; and the houses which

were thrown down by the dreadful explosion of gunpowder
in 1807, in the best part of the town, have not been rebuilt,
their site being converted into a promenade.

We set out after dinner for Haerlem. Among the first
objects of my attention was the organ, and I found out
the organist, who played to me for an hour. It is certainly
worthy of its reputation. I have not had so great a musical
treat for a long time. The organist, Mr. Schumann, plays
remarkably well, and played such pieces of music as were
most calculated to shew its various powers. One piece in
particular was very fine, it was, I may almost say, the
representation of a thunder storm, for it only wanted
lightning to complete the illusion. The "Vox Humana"
I was disappointed in. I asked the organist to play the
hundredth psalm, and certainly I never heard it to greater
advantage. The instrument itself is very handsome; it is
a sad pity that the rest of the church does not correspond
with the magnificence of the organ, for in place of the
beautiful decorations of the Cathedrals of York and Wells,
there are bare whitewashed walls, and the great space of the
nave is filled up with shabby pews, and some hundreds
of old ricketty rush-bottomed chairs, besides the cold
service of the Church is not worthy of such an accompaniment. The organ was begun in April, 1735, and finished
in September, 1738. The statue of Lawrence Koster, the
inventor of printing, was opposite my bedroom window,
and the house where he lived was next door.

Within a quarter of a mile of the town, in the wood
of Haerlem, is a very fine house that belonged to Mr.
Hope, and there used to be a fine collection of pictures
in it. The house was bought by King Louis, and he resided
in it for some time; it is at present uninhabited, but there
still remain the boxes for the sentinels before the gates.
The beginning of May is the time to see Haerlem, for

then there is one of the grandest displays of tulips, hyacinths, &c., which is to be seen anywhere; and this collects a vast number of people from all the adjoining country to see it.

We left Haerlem for Amsterdam in the evening, the road is a straight line nearly the whole way, and the quantity of water on each side of it, at this season, has a very odd appearance.

At the bridge over the narrow communication between the lake of Haerlem and the Y or Ai, the French built a very strong fort, which fifty men, I think, might have kept for a very long time, and completely stopped the communication by this road, but with the usual pusillanimity which appears to have everywhere marked the retreat of the French from Holland, it was soon abandoned. The inhabitants were in great alarm that the French would have cut the Dyke at this place, which they might have done with great ease, and it would have inundated nearly the whole Province of Holland.

I have already told you how much this country has been impoverished during the time it has been subject to France, but there are few things which have surprised me more than the prodigious number of the poor who depend upon charity for a great part of their subsistence.

The great swarms of beggars that I saw in Leyden led me to make particular inquiry there, as to the number of poor who received relief; and I was much astonished indeed to hear that out of a population of twenty-two thousand, no less than eleven thousand have this winter received money, bread, and potatoes. My informer was a very intelligent person, and he told me that he had the best means of knowing the accuracy of his statement. Upon making the same enquiry at Amsterdam, I learned from several sources that the number of persons who have regularly got relief this winter amounts to seventy thousand, the population being about a hundred and eighty thousand. You may easily conceive, therefore, that a very large sum is levied upon the inhabitants. The money is not raised by any direct Poor-rate, but a multitude of indirect taxes, besides that which is paid out of the fund raised by the town upon the inhabitants. The house where I lodge pays eighty florins a year to the town, the rent being six hundred. The number of charitable institutions in Holland has been often remarked.

In Rotterdam there is a Society of some of the most respectable inhabitants, whose duty it is to afford relief by way of lending money to such as are under immediate pressure : small tradespeople and others of the inferior classes, who are ashamed to ask publicly for assistance, and there is great attention paid to affording relief in the most private manner. They also supply different implements, for which they are repaid in small sums weekly. A mangle is a very common thing for them to give, and a great many women now earn a very comfortable livelihood by the mangles which this Society has supplied. It has been established several years, and they have few instances on record where they have not been repaid. I do not find that there are any institutions like our dispensaries, the poor are visited by certain physicians paid by the town ; but there is another excellent plan adopted for this purpose : there are certain societies, formed by the poor themselves, where each person pays to a physician who is appointed by the Society, two-pence a week, for which he attends them during illness, supplies medicines, and provides a decent burial if they die ; thus it is the best interest of the physician to cure them as soon as he possibly can ; the physicians are generally young men, but of the most respectable education. In recompense for the dreadful

sufferings the country has undergone during the dominion
of the French, they have received two great blessings which
probably are worth a great deal more than all the money
they have lost. These are a complete toleration in religion,
and a universal diffusion of instruction among the lower
classes, and having these, if they now obtain the blessing
of a good Government, and are able to profit by a few
years of peace, they have a very bright prospect before
them. There are few countries where religious differences
have been carried to a greater length than in Holland, and
where more mischief has been produced in consequence,
but happily that spirit is almost extinct, and every one
worships God in the way he thinks best without being
subject to any penalty for so doing, either directly or
indirectly, by being disqualified for any public situation.
The present Government have fully recognised this most
wise principle, and in the Council of Amsterdam, lately
appointed, there is both a Roman Catholic and a Jew. I
had the pleasure of seeing three very good collections of
pictures, viz., the public museum at the Stad-haus,
and two private collections belonging to Mr. Van Winter
and Mr. Coesveld. The collection at the Stad-haus, or
Palace, as it is now called, is now almost entirely of the
Dutch and Flemish schools, and contains many valuable
pictures; those I was most pleased with were the works
of Wouwerman, Paul Potter, Buchuizen, and Vander Velde.
There is also a wonderful picture by Gerard Douw, where
the different lights and shadows occasioned by four or five
candles in different parts of a room, are represented with
extraordinary skill. The greater part of this collection was
made by King Louis. In gratitude, I suppose, they suffer a
full length portrait of him to remain in the great hall. But
indeed he seems to have been very much liked by the Dutch,
they all seem to speak well of him. He was, however, daily

more fond of getting all the power into his own hands,
although when he began to reign, he expressed a great
anxiety to give his subjects as much freedom as possible.
Some time before he was turned off, he had begun a
museum of natural history, the formation of which he
entrusted to Professor Rainward. The greater part of what
was good was removed to Paris, when this country was
annexed to France. Mr. Coesveld's collection is principally
remarkable in containing the greatest number of Spanish
pictures that probably exist anywhere together out of Spain.
He has specimens of fifty-four different masters; they were
collected by himself during a residence of four years in
Spain when the Revolution first broke out. He has also
some very fine pictures of the Italian School; he has a set of
thirty-six sketches in bistre, by Murillo, representing the
history of the Virgin, which are very wonderful productions.
The classical library of the Professor Van Lennep is considered as the finest in Holland. He has many books
which are seen with great pleasure even by those who are
not able to appreciate their real value. I was much pleased
to see an edition of " Aubus Gallius," which had belonged to
Erasmus, containing his handwriting in many places; and
also a copy of " Theocritus," that had belonged to Grotius.

I dined with Mr. Van Lennep (having received from Mr.
Falck a letter of introduction to him and Mr. Rainward)
and met at his house Professor Cras, a fine old man of
seventy-six, full of activity both in mind and body. He is a
Professor of Civil Law, and very much looked up to by
the most intelligent people of Amsterdam. When the
Academy of Sciences at Stockholm some years ago proposed
a premium for the best eulogium on Grotius, Mr. Cras
carried off the prize. He was so kind as to present me with
a copy of it.

Boomte, *March 23rd*, 1814.

I write from a small village between Osnabruck and
Diepkolz, on the road to Bremen. We left Amsterdam
last Saturday morning, the 17th March, after an early
breakfast, and bent our course towards Utrecht; the direct
road would have been by Naardon, but that being in possession of the French, we were obliged to avoid it. The road
for some miles is by the river Amstel which near the town
is very wide. It was completely frozen over, and a vast
number of peasants were upon it, pushing small sledges,
which carried a little sail before them, containing various
things for the market of Amsterdam. We reached Utrecht
at one o'clock. Within a few miles of the town the road is
very pretty, passing by the side of the river Vecht, which,
strange to say, shewed hardly any signs of frost, although
we had left a severe winter a few miles behind us. The
Rhine is a dirty canal here. There are the ruins of a fine
Gothic Cathedral, which is well worth seeing; the great
nave is gone, but the tower, which was at the west end of
it, is still standing, and the choir is occupied as a church.

In the afternoon we set out for Amersfoort, the road passes
through some pretty country, well clothed with trees, and
near Amersfoort some sand hills relieved me for a while
from the monotonous horizontal watery space I had been
for some time accustomed to. It was Sunday morning, and
the peasantry were going in considerable numbers to a village
church a few miles from Amersfoort. Their costume and
general appearance was quite different from any that I had
yet seen, particularly the men, who had large coats dangling
down to their heels, with an infinite number of buttons, and
the button-holes worked with worsted lace, they were all
the same colour—a dark grey—the buttons and lace black.
They all wore a very large slouch hat—the shovel hats of our

clergy—the boys were dressed exactly like the oldest men, so
that when their backs only were seen, there was nothing to
lead you to guess them under sixty years of age ; they had
all wooden shoes—men, women, and children; even the
parson himself, who appeared among the last, was shod in
the same way.

Near Apeldoorn we passed by Loo, the Palace 'built by
William the Third, and where he used to live very much. I
cannot admire his taste much, for it stands in the middle
of a wild heath, and has only a few trees immediately round
it. It was, however, put in complete repair, and furnished
by King Louis, who appears to have had the same taste as
William, for he resided three summers there, and Buonaparte
and his Empress stayed three days there about a year and a
half ago. At Apeldoorn we bade adieu to good roads; from
Amsterdam to that place nothing can be better, it is excellently well paved, with a good covering of earth to render
it easy. This was made by the French, another of the
many good things which the French have done for Holland,
and those countries which they have adopted', and certainly
they have afforded some compensation for the evils they
brought with them; one of the most considerable is the
excellent roads they made, the Dutch themselves acknowledge
it with gratitude.

But in speaking of bad roads I need not confine myself
to that between Apeldoorn and Zutphen, but say that the
whole way from thence to Osnabruck exceeds in badness
anything that I had any conception of. In general it can
hardly be said that there is any road, and the ocean of
heath we passed over, differs only from the watery oceans
in being a pathless waste with occasionally some traces of
wheels. To tell you how deep our wheels sunk in mud and
water, the depths of the ruts or the size of the stones, will
not give you so good an idea of these roads, (as they are

misnamed) as to tell you with a light carriage and three
horses we did not exceed, on an average, three miles and a
half in the hour. The wooden bridge at Zutphen over the
Yssel, is so constructed, that when they fear an inundation
it can be removed by means of boats, and placed in safety,
which is the case at this moment, and we had, therefore, to
be ferried across the river. We arrived just in time before
the shutting of the gates, which takes place at six o'clock.
Zutphen is garrisoned just now by some hundreds of the
Prussian Landsturm. The vicinity of the French at Deventer
renders it necessary for them to keep a good look-out. We
found a tolerable inn, and next morning as soon as the
gates were opened we set out for Delden. If there is anything
remarkable to be seen at Zutphen I did not hear of it. It
differs only from the other towns of Holland I had seen in
being less clean, and having no canals in the streets. I
think that the cleanliness for which the Dutch are so remarkable is very much confined to the Province of Holland itself.
From Zutphen our way lay by Lochem and Goor to Delden,
and had a long dreary stage next morning to Bentheim,
within a few miles of which I saw the first rook I have met
with on this side of the water, and I felt some pleasure in
seeing my old friends again. Bentheim is finely situated on
a high ridge of rock, which rises above the plain which sur-
rounds it on all sides about three hundred feet. The highest
part is occupied by a very large castle, which bears marks
of great antiquity. It stands on a very fine situation, the
rock is a sandstone, which frequently breaks out into great
blocks, producing by their various massive groups a very
picturesque effect ; as we advanced the villages became more
and more dirty, and the appearance of the people more
miserable and poor. The landlord of the inn, a very old
man, showed us the room in his house where George the
Second slept on his way to Hanover. I am afraid his

Majesty passed but an indifferent night, if his host had no
cleaner notions in his younger days than he appears to
have now.

Our next stage was Rheine, over heaths and morasses, the
melting snow rendering every place worse than its natural
bad state. Osnabruck is a curious old town. There are two
cathedrals, and as the town is half Catholic and half Protestant, each has a cathedral. The Lutherans occupy the most
modern, which is rather better worth seeing than most of the
churches of the same kind in Holland. The other cathedral
is a very curious pile, and was built, as I was told, in the
reign of Charlemagne; the interior is not fitted up with any
splendour, the decorations of the altars being very tawdry.

The great road which the French had begun to make
between Hamburg and Wesel, passes by Bremen and Osnabruck, and a great part of it is finished between these two
places; it is a very fine, broad, paved road, and was a great
luxury after the jolting we had endured for some time. We
hope to reach Bremen to-morrow evening; and thus we have
come to the end of our journey in six days, having come
about three hundred English miles. From Bentheim we
had travelled through a great deal of country belonging to
the Electorate of Hanover, and almost every little inn had
the sign of George III. Rex, which, from the freshness of the
paint, shewed they had been there when the French were in
the country. The inhabitants in the German part of the
road have the appearance of the most wretched poverty,
both in the squalid appearance of their countenance, the
rags that cover them, and their miserably dirty hovels.
Among some of the peasantry about Osnabruck, the most
remarkable part of their costume is the head-dress of the
women, which is a small cap of silk or cloth, embroidered
with gold or silver, stuck on the back part of their head,
with a piece of lace or muslin standing erect from the

edge of it, with their hair combed smooth back from the forehead, very like the glory which you see painted round the heads of some saints. Were they in other respects well dressed, and if their face and hair did not exhibit too strong proofs of a scarcity of soap, I daresay this head-dress would be becoming, but except in Osnabruck, where I saw some pretty women, I have seen nothing but the most ordinary countenances, both in the women and the men. In my next I shall tell you what I think of Bremen.

Bremen is built upon each side of the Weser, but the principal part of the town is on the north bank. It is of an oval form, and fortified, but quite incapable of defence, except for a very short time. The river is about half the breadth of the Thames, rapid, but so shallow that no larger ships than lighters can come up to the town, all the merchant vessels discharge at Elsfleth, a small port belonging to the Duke of Oldenburg, and to whom a duty is paid. It is about one per cent. on the value of the goods, so that if the trade of Bremen becomes great, as it is likely to do, the Duke will derive a very considerable revenue from it.

The great church called the Dom is a vast pile, excessively ugly, and filled from one end to the other with pews and galleries, adorned with all the most distinguished characters of the Old and New Testaments, painted not by a Raphael, certainly. Under this church there is a very remarkable vault called the *Blei-keller*, where animal substances do not putrify, but dry up and become in the state of mummies without being embalmed, or any other preparation. There are about a dozen bodies exhibited in their coffins, several of which have been there for above a hundred and fifty-years. There is one English Countess, of the name of Stanhope I think, and a Swedish general, the flesh is dried, and becomes of a pale brown colour, but the features are preserved; they are extremely light, their appearance is more like the fungus called amadou, which the Germans use for lighting their pipes, than anything else. What this phenomenon is owing to, I cannot conceive; there is a very free circulation of air in the vault, but that is, of course, insufficient to account for it.

The most remarkable place which is shewn to strangers in Bremen is another cellar, but it is one of a better description. It is under the Town House, and is filled with old hock. I was at a supper which Mr. Kulenkamp gave in the cellar, and we visited the various vaults. One contains twelve large casks, which are called the twelve Apostles. The finest wine is contained in a vault called the Rose, where we tasted hock of 1624, and a wonderfully good wine it is. This old wine, which belongs to the town, is not sold, but is given gratis to the sick upon the simple order of one of the physicians of the town. The finest sort is kept as a bon-bouche for Princes and Grandees when they visit the town, who are generally presented with a few bottles of it. Presents used often to be sent to our old king, who, it seems, is very fond of it. I am not surprised at his Majesty's taste. The people of Bremen are quite surprised that the French did not carry off this great treasure of their town, they can find no other reason but that the French do not like this kind of wine. I went to the theatre, a very shabby, dirty place; the play was " Mary, Queen of Scots," one of Schiller's. Had I arrived a day sooner I should have heard Madame Becker, who is considered the finest singer in Germany.

Trade is the only thing thought of here, and the merchants are, in general, wholly devoted to their counting-houses, and all the time which is not spent at the desk is passed in their clubs, like those in Holland, where they smoke, drink, and play cards or billiards. I had the pleasure of meeting Mr. Charles Parish of Hamburg here, I had a letter for his house from Mr. Baring; he has been in Bremen since the month of November, constantly expecting the fall of

Hamburg. One day when I was dining with him we had an alarm which for some hours gave us a little uneasiness. The Russian General, Clott (and a clod of a fellow he is), who is Governor of the town, was dining with us, and just before he came, he received a dispatch by the Estafette, from the Duke of Cambridge at Hanover, stating that Davoust, who commands at Hamburg, had quitted the town, with his whole troops passed over to Harbourg, and was in full march upon Holland with fifteen thousand men, by way of Bremen. As two days had elapsed between the time when Davoust was understood to have left Hamburg, and the arrival of the Estafette, we concluded that he must be within a few hours' march of the town. The General had sent out his Cossacks upon the roads, and sent orders to some Danish troops which had marched towards the Rhine the day before to return. The senate was assembled, and the result of their deliberation was to defend the town. There was about eight thousand troops, besides the Burgher Guard, which it was thought would keep off Davoust, until General Beningsen could come up behind him. I confess I did not much relish the idea of a battle, and I wished myself a little further from the scene of action. However, in the morning we were relieved from all anxiety, by the arrival of' a person from near Hamburg, who said that the French had made a sortie, but had returned to the town. The accounts of the sufferings of Hamburg are quite dreadful—every day people are turned out of the town, and they are turned out destitute of everything. Bremen just now supports about eight thousand of these poor people.

We have no idea in England of the numberless oppressions which those countries have borne, which have been the seat of war. There are few of the private families of Bremen who have for some years past, been without soldiers quartered upon them.

Mr. Kulenkamp, who is one of the principal inhabitants, has generally had the General, whom he is obliged to receive as an inmate of his family. The Russian General has been in his house for four months, and he has five other officers besides, quartered upon him. He reckons the expenses he is put to by the officers who are quartered upon him, at not less than five hundred pounds a year. But the expense is not the worst part of the grievance, for the privacy of his family is completely destroyed, and in the summer time when he takes his family into the country, his wife is obliged to come three times a week to town to look after their house, to see that everything does not go to ruin. There were a great many different troops while I was there, the greater proportion of whom were Danes, who were on their march to join the grand army in France. They were by no means good-looking troops; they were of all sorts and sizes, badly drilled and badly clothed. There were about three hundred Cossacks, whose appearance quite corresponds to the nature of their warfare, for they are more like ruffian banditti than any thing else. The Hanseatic Legion there were very fine troops indeed.

Bremen has flourished a good deal by the contraband trade, which has been carried on to so great an extent from Heligoland; and from the great destruction of property at Hamburg it is likely to become a much more considerable place than it ever was before. In the article of coffee alone, there were received from Heligoland, between last November and the beginning of January, four millions of pounds weight, and there was more in the market when I was there. Its chief exports are linen and corn.

A few days after my arrival at Bremen the frost broke up, and the ice in the river gave way, it was a fine sight to see the large islands of ice which floated down the stream.

But the melting of the snow on the higher country so
swelled the stream, that it overflowed its banks and
inundated all the adjacent country. Those houses in
Bremen which' were nearest the river had their lowest
stories filled with water. On our return to Holland we
passed not far from Deventer, which is still in the hands of
the French; the evening before we passed, a heavy bombard-
ment had taken place, and I met eight heavy pieces of
ordnance going against it. They had come from Gorcum,
where they had been employed in the siege. Having never
seen cannon passing along a road, except when going to
a parade, I felt a very unpleasant sensation at seeing
these instruments of death slowly moving along. I had
been unfortunate in being absent from Amsterdam at the
time when the ceremony of swearing in the Prince took
place. I understand it was a very fine spectacle, and
exceedingly well-conducted.

In the Trekschuyt, between Amsterdam and Gouda.

April 12th, 1814.

In order to have seen as many of those things which
are peculiar to the country in which I now am, as my
opportunities will allow, I have chosen this mode of conveyance. A few days ago I engaged the roof, which is a
small cabin about six feet by five, in the stern of the boat,
with a seat on each side covered with large crimson velvet
cushions, on which I can sleep if I please; a small table on
which I now write, and a wax candle, which I bought in
Amsterdam, in a bright brass candlestick. I embarked at
eight o'clock, and I am told that at five o'clock to-morrow
morning I shall arrive at Gouda. In a small cabin next
to me, about the same dimensions as my own, there are
some voices talking Dutch, and farther on there is a larger

place containing a promiscuous company, among whom I
hear two English voices considerably above the rest; the
usual accompaniment of the pipe is in the mouth of almost
everyone in the general cabin, and by the time of our arrival
at Gouda, the atmosphere of that place will be in a fit state
for chemical examination. The motion of the boat is
quite insensible, and I write as easily as if I were in a
room.

I have, in my former letters, told you how much the
people of Holland suffered in their property while they were
under the dominion of the French, but that grievance was
slight in comparison of the oppression which they endured
from the conscription, and the tyranny of the public officers.
Every *Prefet* and *sous prefet* was a tyrant, acting entirely
according to his own humours without restraint. If any
complaint or remonstrance was made that they were not
acting according to law, their usual answer was that the
laws were not intended for them; they had their instructions
from Paris, and that was enough. These fellows had full
power to send anyone to prison without stating any reason.
The Comissaire de Police at Rotterdam appears to have been
one of those who exercised his power to the fullest extent;
as a refinement of cruelty he sent people to the madhouse,
in place of the common prison, where, in addition to the
misery of confinement, they had that of being in the same
place with people in the worst state of madness. One gentleman in the town of the first respectability was sent there,
after having all his papers seized, in consequence of having
been suspected of holding an intercourse with England; he
was kept there for three days on bread and water, and was
only released in consequence of several of the principal
inhabitants coming forward and offering every security for
him that might be required.

The conscription appears to have caused more misery than

any other of the forms of oppression they laboured under. When
the country was annexed to France the conscription was first
extended to it, but Bonaparte considered that they ought to
have furnished supplies to his armies sooner, and the first
year he ordered the conscription to be raised for three
preceding years. I have not yet ascertained the exact
number which the land furnished annually, but I believe
it was not less than five thousand men, the age being from
twenty to twenty-six. The expense of a substitute was so
great that it put it out of the power of many most respectable
people to procure one for their sons, from five thousand to
eight thousand florins were the common prices, which is
from five hundred to eight hundred pounds ; and the substitute having to be approved by the *Prefet*, he had an
opportunity of exercising his resentment against particular
people by refusing to admit the substitute they had procured.
When Bonaparte's army was destroyed in Russia, we heard,
in England, a great deal from the French papers of the zeal
of the different departments in making voluntary levies of
horses. These voluntary donations were brought forward in
this way, the Prefect wrote to the mayors of the different
subdivisions of the departments, that they must *vote* a certain
number of horses, which the inhabitants of their jurisdiction
must supply, and make the return to him on a certain day.
These returns the *Prefects* sent to the proper officer at Paris,
and they were published with a fulsome address of the *Prefect*,
as the proofs of the patriotic feelings that pervaded the
nation. But very soon after this measure there came a
more dreadful scourge, the levying of the *Gardes d'honneur*.
Bonaparte, wishing to have hostages of the principal families
throughout his dominions, fell upon this dreadful scheme ;
ten thousand were to be raised throughout the empire, and
they were called *Gardes d'honneur*, as they were intended
to protect his sacred person. They were raised in this way.

The Prefect of each department sent a list of the sons of all the principal families in his department to Paris, and when it was returned with the names who were selected, he sent an order to these young men to repair on a certain day to a place of rendezvous, and that they must equip themselves with uniform, arms, and a horse. You can easily conceive the horror with which such an order was received in many families. No substitute was allowed, not even bad health was admitted as an excuse; one gentleman who came to the Prefect to say that it was impossible for his son to serve as he had a complaint which prevented him from riding on horseback, was answered, "then he must walk by his horse's side." I became very well acquainted with one who had served; Mr. Borsks, of Amsterdam, a son of one of the most opulent merchants. His father offered to equip twenty-five men in his stead, but was refused; the young man was marched off, and his father died soon after; his death, it is generally said, was very much owing to the grief he suffered for the loss of his son. The young man was fortunately taken prisoner by the Austrians a short time before the battle of Leipsic, and was soon released. Many who obstinately refused to go were dragged from their families by *Gens d'Armes* to the place of rendezvous, and were held by force while they were measured for their uniforms; they were marched to Mayence, and several died of fatigue on the road. They were chiefly sent to places far remote from their homes, for they were not kept round Napoleon's person, as they were told they were to be. M. Suesmondt, of Rotterdam while I was there, learned that his son was at Grenoble. What was not the least part of the oppression was the insult to their feelings that accompanied this levy, for they saw in the *Moniteur* that they had come forward with enthusiastic zeal to guard the sacred person of the great Napoleon.

E

Pillaged and oppressed in so many ways you may easily
imagine that the country was ripe for revolt, whenever the
opportunity should present itself; but whether it is that
their sufferings had reduced them to a state of imbecility, or
that their natural disposition renders them incapable of
exertion, from all I can learn, it appears quite clear that had
not the French shewn a most extraordinary degree of
pusillanimity, the revolution would not have taken place at
the time it did; it is pretty clear that a plan had been
formed by some of those persons now in the Government,
who were living in a retired way and had been doing so for
some years, to recall the Prince of Orange; but it was so
far imperfect that they were much alarmed at the signals of
revolt appearing so soon as they did. When the accounts
came of the advance of the Allies, the Corps of Douaniers at
Amsterdam took fright, and without beat of drum, made
their escape in the middle of the night. Next morning there
was much bustle, and the cry of Oranje Boven, and the
appearance of orange cockades were signals of commotion.
Very soon after the wooden houses where the Douaniers
used to sit and collect the duties were brought together, and
committed to the flames; this tumult very quickly spread
and communicated itself to the neighbouring towns, where it
appears that a few resolute individuals came forward and
took a very decided part; among these a Mr. Kemper, one
of the Professors at Leyden, took a very distinguished part,
and Count Steerum with his two sons had the boldness to
walk through the Hague with orange ribbons on their hats
while the French were still in the place. Very soon
afterwards the Prince was sent for, and at the same time
a messenger was sent to the English Admiral cruising off
the coast for assistance. Although there was a very general
expression of feeling among the people, there was a very
anxious interval between this demonstration of their

intention and the first arrival of the Marines sent on shore
by the English Admiral. Nothing can mark it more, than
the manner in which Mr. Grant was received; an adventurer
who on the first receipt of the news of the revolution had
set out from London on a mercantile speculation, and came
over in a fishing boat; he had brought with him a
volunteer's uniform, which at the request of some of the
principal people he put on, and paraded the streets in it. I
have been told by several people that it is wonderful the
effect which the appearance of the scarlet jacket produced.
He was considered as an English Officer, *the advanced guard*
of reinforcements from England, and he was not only the
means of spreading the flame with greater rapidity, but
caused great alarm to the French, who were under the same
impression as the natives. This terror of the French was
shewn in almost every place where they were. At Amersfoort
three Cossacks entered at one gate, and 400 French fled out
at the other. I suppose there never was an instance of the
government of a country being overturned in so short a time
and with so little bustle, as the late revolution in Holland.
It is like a dream to the inhabitants themselves, and I
fancy they are still in a great degree under that impression,
otherwise they would be more active in raising a defence for
themselves than they are. The dreadful msssacre at Bergen
was entirely owing to the dull stupidity of the inhabitants,
who went quietly to bed without raising their drawbridges,
and the French entered in the night without opposition.
When the Prince landed at Scheveling, the good people
asked him what they should do with him; he, not knowing
what situation affairs were in, could not advise them, so they
thought it best to take him to the minister's, who gave him
a dish of tea, and there he waited till the people of the
Hague came to meet him. That matters did not go on with
greater activity, there is this to be said in favour of the

Dutch, that they were totally without arms or ammunition, the French having carried off everything of the kind. The first people that assembled in Amsterdam held no better arms than muskets without locks, old swords and pistols, and anything offensive that they could lay their hands on. But they might ere this have had a very tolerable force in the field, had there not been a great inactivity on the part of government. The Landsturm at Amsterdam are armed with pikes which they got for the first time about three weeks ago, while I was there, and a very odd looking assemblage it is.

[Mr. L. Horner returned home in April and in July his second daughter was born.* In March of the following year he went to reside in Edinburgh. The removal from their many English friends was a great trial both to him and his wife. To her it was a separation from her nearest relatives; but they never hesitated in taking any step, however painful, which they considered was their duty, and they had only one mind between them on every subject.]

* Afterwards married to Sir Charles Bunbury, Bart.

CHAPTER IV.

1815.—October 1816.

To his Sister.

London, *February* 25*th*, 1815.

My Dear Fanny,—We were at a party last night at Lady Romilly's where we met Sir James Mackintosh, Mr. Sharp, and some others, who all came up to me to give me joy on the very distinguished appearances Frank had made this week. Sir James Mackintosh said that he was sure of a great deal, and he expected more, but his highest expectations were greatly exceeded. He had talked with many old members, and particularly Whitbread and Lord Morpeth, and they both admitted very decidedly, that both speeches were such as would have been remarkable in the best time of the House of Commons. He said that no man could be more completely in possession of his audience than he was, and during his speech on the Corn Bill he was listened to with the most profound attention by his most zealous adversaries. They consider that by these speeches he has most completely established his rank in the country.

From Dr. Marcet.

London, *April*, 18*th*, 1815.

My Dear Sir,—You ask me what our *sound heads* think of the question of peace and war, and of the tremendous convulsion which has just taken place. It is not easy to answer such questions in a few words; but I can safely say that all the soundest heads I know, with the exception of our friend Blake, and one or two other high-minded Britons, are for giving peace to Buonaparte upon the basis

of the Paris treaty, and trying once more, not his sincerity,
for there is no such thing in diplomacy, but his perseverance
in a new line of policy, which his own interest has induced
him to adopt. You know how heartily I detest the man,
and how little I like the French, and therefore you will
give me credit for not being biassed by undue partialities
in this question. But I confess that the idea of beginning a
new and tremendous war, connected, tacitly if not openly,
with the restoration of a dynasty which France is determined
to reject, appears to me a desperate measure. It is no less
than realizing and securing at once war and all its evils, for
fear (in case we continued at peace) war might be, after all,
the ultimate event; as if delaying the Income Tax, and the
waste of blood and treasure, with a chance of ultimately
escaping those evils, was no object whatever!

I know that the reply is, that Buonaparte would in the
meantime prepare himself, but should we not see these
preparations and keep pace with them conjointly with our
Allies? And would not Buonaparte (whose *present* wish for
peace nobody questions) consent to some stipulations that
would afford some kind of pledge of his future conduct.

However, I much fear there is now very little hope left,
and that we must soon begin to fight, *bongré malgré*, against
twenty-five millions of people, who have more than once
subjugated half of Europe, and yet are ready to consent to
almost any terms you might propose to them. If Buonaparte
should be destroyed by this new coalition, I shall thank the
stars of Europe that the voice of prudence had not prevailed,
but I shall not be the less persuaded that blind passion has
been the chief agent of this new war, and that the Emperors
and Kings, and most of their subjects, have shown more
temerity and spirit of revenge, than regard to the welfare
and happiness of the world.

Our accounts from Switzerland and Geneva are as favour-

able as the situation of affairs can admit of. The Swiss are
certainly disposed and determined to side with the Allies, if
they cannot preserve their neutrality, but strong hopes are
entertained that both parties will agree to suffer the Helvetic
league to remain spectators. The Congress have given to
Geneva a pretty little arrondissement of territory, which
includes Salève, a mountain which would laugh at Ben
Lomond.

Pray remember me and Mrs. Marcet most kindly to Mrs.
Horner, and to your father's family,

And believe me,
Ever most sincerely yours,
ALEX. MARCET.

———

To Dr. Marcet.

Edinburgh, June 16th, 1815.

MY DEAR SIR,—As I hope to have the pleasure of seeing
you again in less than three weeks, I might defer writing to
you, were it not that if I should leave your last letter
unanswered, you might think that I do not sufficiently value
a correspondence, which I assure you will be a source of
great gratification to me in my exile. We have succeeded in
getting a house entirely to our mind within a short mile of
Edinburgh, and my time has been incessantly occupied in
the tedious occupation of preparing it for our reception.
The house is very comfortable, and we have a garden of
nearly an acre in extent. The prospect you give us, of
you and Mrs. Marcet visiting Scotland this autumn, is
very pleasant, and we depend upon your making our house
your home, so long as it suits your convenience. We
have a good bed for you, and a hearty welcome will, I trust,
make up for the homeliness of the cheer. Mary will be

delighted to see her friend Louisa, and to practise together
Monsieur Olivier's steps. . . . .

Greenough and Warburton were so good as to give me
some information about the Geological Society lately, the
loss of which I feel greatly, as indeed I do that of all the
good fellows I was in the habit of meeting there; but I must
now be contented with an annual glimpse of them. I have
not yet seen any one here who will be likely to make up in
any degree for the loss; the habits and style of society are
very different indeed. Has Davy returned? In what state is
he, does he talk much of Popes and Princes? What does he
say of his friend Murat, who, by-the-bye, has gone off the
stage very clumsily? War seems to be the mad determination of our Government, and even the wish of the country,
although if it continues, great calamities must befal us, and if
we are to judge of the spirit of the French nation by their
newspapers, the contest will not be soon terminated. I
cannot see the question in any other light than that war is a
scandalous attack upon the liberties of France, and as such I
most heartily wish that the Allies may fail in their attack, if
they are so mad as to make one. It seems very probable
that the friends of national liberty are much stronger in
France than they have ever been before. I
shall expect to find you much advanced in your mineralogical
studies. I have been laying aside some little bits for you. I
shall not be long in town without finding you out, and I
believe I still remember accurately your hospital days, so
that I shall take my old time of calling. . . . . . .

I am always very faithfully yours,

LEONARD HORNER.

[In the month of July Leonard went up to London on
business; his brother Frank was on the Circuit for two or
three weeks after his arrival, and during his absence Leonard

undertook the troublesome task of sending the books and furniture from Francis's chambers in Lincoln's Inn, to a house he had taken in Great Russell Street, as Frank was very ill suited to manage these matters for himself, and when he returned from the circuit, he thus found his house comfortably arranged.]

From Francis Horner.

Exeter, *July 17th*, 1815.

MY DEAR LEONARD,—On coming here to-day, I found two letters from you, the last of them announcing your arrival in London. Since your departure from Edinburgh, I have accounts of all being quite well.

I read in the proof-sheet as much of Whishaw's account of poor Tennant * as was printed for the first part; it appeared to me extremely good, executed with all his usual candour, and sufficiently characteristic in the view given of his eccentric, amiable friend.

I have been much incommoded since I left town by an attack of my old complaint. I mention it lest you should hear by accident of my being unwell. This is very trouble-some, but I am assured that I shall in time get rid of it altogether. Travelling is not good for it, and hot weather worse. But do not imagine me at all seriously affected at present. I have scarcely been impeded in doing business as usual, and the quiet, confined life to which I have restricted myself for the rest of the day, when I am not in Court, has given me an opportunity of reading a good deal, which has delighted me much; my late London life affording so few opportunities for anything like the regular perusal of a book quite through. Since I left town I have read Xenophon's

* Smithson Tennant, Professor of Chemistry at Cambridge, killed by a fall from his horse at Boulogne, February 22nd, 1815.

" Anabasis " in the original, from beginning to end; the
longest piece of Greek I have ever got to the end of in
a regular way.

Most affectionately yours,

FRA. HORNER.

From Francis Horner.

Exeter, *July* 26th, 1815.

MY DEAR LEONARD,—I shall be very much disappointed
indeed, if we do not make out the journey to Scotland
together, which seems to me very practicable; provided you
will stay with me a little time in London after my return
from the Circuit. With respect to the places you wish to go
to, we can easily plan a route that will include them all.
While you are gone to Bristol and the Quantocks, I shall not
be sorry to pass a day or two in Gloucestershire, and you can
pick me up at Malvern, where perhaps I shall go to see the
Abercrombys, or at Cheltenham, where I should like to make
you acquainted with the Carnegies. I will then go with you
to Birmingham for the time you may require. After that I
have a wish to pay a visit at Etruria, to my friends the
Wedgewoods, on purpose to see their manufactory, as well
as to see them. I have said that you will accompany me,
it is not two miles off the road. Newcastle, to which you
must go, is a place where I shall have no objection to stop a
day or two; and I believe we may from that still keep the
promise which you told me you made to Lady Anna Maria
Elliot, of calling with me at Minto. There is a new road,
I believe by the Cheviots, which would take us towards
Edinburgh in that direction without any increase of distance.
In the outline, therefore, the plan seems both practicable
and pleasant; we will settle the particulars when we meet,
only you must arrange I think, amusements or employment

for a fortnight with me in London. I can hardly do with less. I am very much obliged to you for the trouble you are taking about my books. Do not suffer it to interfere inconveniently with your occupations.

Of course you have heard by this time of Buonaparte being in Torbay. People from Torquay and all the watering places round, go out in boats, and crowd round the Bellerophon, to get a sight of him as he walks the deck. The stories here are that he often bows very graciously to these gazers, and is dressed in his old fashion—a green frock coat, with a star, and scarlet facings.

Yours ever affectionately,

FRA. HORNER.

P.S.—I continue to mend, and feel more at ease; and the great point of all is gained, I now get some tolerable sleep.

From Mrs. Marcet to Mrs. L. Horner.

Clapham Common, *October 14th*, 1815.

MY DEAR MRS. HORNER,—On my return from the Continent I received a letter from you. . . . . We spent full three months on the Continent, having left England on the 7th of June, and since we have been so fortunate as to escape, not only all danger, but even inconvenience, we certainly could not have pitched upon three more interesting months to travel in. Then if fate had given us the direction of events, they could not have occurred more opportunely. The Allies waited our departure from Brussels, and gave battle to Buonaparte as soon as we were out of sound of the cannon. The French were skirmishing in the environs of Geneva, till the very day of our arrival, and it was on that day only, that the arrival of 55,000 Austrians relieved the town from the apprehension of an attack from the French,

who had been in great force in the neighbourhood. Then
at Paris, scarcely a painting was taken down from the walls
of the Louvre, and not a single statue removed till we had
visited the gallery; this is a piece of good fortune at which
I shall ever rejoice; nor can I cease to regret the time we
spent in shopping and visiting, or indeed in passing our
mornings anywhere but in the divine Gallery which no
longer exists!

We spent two months at Geneva, or in making excursions
in Switzerland; it was delightful to see the Doctor
surrounded by the friends of his early youth in his native
country, a country which at his last visit had not existed,
and which has been almost miraculously restored and
preserved; the spirit of military ardour, the enthusiasm of
joy, the burst of patriotism, all produced such an effect on
my mind, that I almost thought myself a Genevese, and I
found it necessary to call to mind that I had left a father
and children in my *own* country. I was particularly
interested at Geneva with the manner in which the wounded
Austrians and French were treated; there was a very
dreadful siege of a small fort, Le fort de L'Ecluse in the
Jura, at a short distance from Geneva; we heard every
cannon that was fired, and to one who has never heard the
sound of cannon but at a review or in celebration of some
festival, you cannot imagine how awful is the impression it
makes when fired in real warfare. During the eight days
this siege lasted, the wounded were daily sent in carts to
Geneva; you would have imagined that the whole town had
no other concern than to assist and relieve them; at every
evening party the ladies were employed in making lint, in
the mornings many of them personally attended the Hospital,
and women, as lovely in person as in mind, were seen not
only superintending the nurses, but themselves actually
attending the sick. All the towns on the borders of the

lake followed the example of Geneva, and some even surpassed her; this was the more to their credit, as the *people* in the Pays de Vaud are certainly Buonapartists, and were extremely averse to the invasion of France by the Allies. However, neither Geneva nor the Vaudois made any difference in their treatment of Austrians or French, they were no sooner wounded than they considered them as brethren.

Our return through France was extremely interesting; in the course of our travels we had seen many sovereigns and great men, and heard their political opinions, but nothing appeared to me so desirable as to learn the sentiments of the PEOPLE of France, and to be eye witness of their actual state. It was a very great gratification to us to find them much better off than we expected; we had apprehended the painful scenes we might witness in a country which had so long been suffering under the scourge of Revolution, and warfare; but excepting one village, which had been burnt on the first entrance of the Allies into France, we witnessed no marks of outrage; the fields were well cultivated, the villages full of people of a cleanly and healthy appearance; we were particularly struck with the rosy, fat cheeks of the children; of squalid poverty we saw no marks till north of Paris, where the shoals of rich and charitable English travellers seem already to have bred up a set of beggars; in the other parts of France, they have disappeared with the convents, which afforded them a miserable maintenance with those oppressive and unequal laws which reduced them to distress. But alas, poverty and distress will I fear but too soon again revisit poor France. The people made no complaints of the troops except of their voracious appetites; they were mortified and humbled, and were almost universally Buonapartists, and I do not think that the spoliation of the Gallery of the Louvre (however equitable) and the treaty of peace, are calculated to reconcile them either to the Allies or the Bourbons. But there would be no end to this subject, and I must say a few words to you of our family. Our return was much embittered by the sickly state in which we found it, and which had been so softened to us, that we expected to find them all pretty well. It was indeed a melancholy contrast with all the amusement and gaiety we had enjoyed, and I almost felt remorse at having been so happy, separated from my family, but it was not merely pleasure we sought in our visit to Geneva, it was a duty we owed to the Doctor's mother and his family, and if we were happy in fulfilling that duty, one ought not to regret it.

We had renounced all thoughts of taking any of our children with us, when we decided on going through Germany, not thinking it right to expose them to the inconveniences, and even danger, we were liable to encounter, in travelling through such numerous armies, and I assure you it excited no small degree of admiration in us to see such immense and numerous bodies of troops, marching in so peaceable and orderly a manner, and seldom venturing to pull a potato out of the ground as they passed along, so excellent was the discipline preserved. I speak not of that of the Prussians after they entered France, we saw nothing of them, but the accounts we heard certainly did not correspond with the conduct of the troops we saw on our route, who were Austrians and Bavarians, etc.

I perceive that I have jumped from the sickly state of the family to the Austrians and Bavarians, without having informed you that we are now all well again, and my father in as tolerable health as he has been for a long time.

We regret extremely to have been absent during Mr. Leonard Horner's visit to London, and beg to be very kindly remembered to him. The Doctor will answer his letter as soon as the first press of business is over, but at present he

is so much engaged with his patients and his lectures, that
he has not a moment to spare. We were very happy to hear
that you are at length settled so much to your satisfaction.
Adieu, my dear Mrs. Horner.

Believe me, very truly yours,

J. MARCET.

P.S.—We visited poor Mr. Tennant's grave on our return
home, and saw the place where he fell, the surgeon declares
that he was perfectly insensible from that moment.

From Francis Horner to Mrs. Leonard Horner.

Bowood, *October 29th,* 1815.

MY DEAR ANNE,—I have spent a most agreeable week here,
in spite of unsettled weather, and leave the place with much
regret. I am going to-day part of my way to Bulstrode, in
company with the Abercrombys, who are going to London.
I shall be settled there myself on Wednesday. It is my first
winter without having Gower Street* to go to upon my arrival.
I hear from town to-day, that Admiral Fleming has received
a letter from Sir George Cockburn, dated off Madeira, who
reports Buonaparte as having been in good spirits during
the voyage, but lethargic, which prevents him from much
reading or writing. He sleeps fourteen out of the twenty-
four hours. But he has taken a great deal to card-playing,
of which he was ignorant when he left Plymouth, but now he
beats everybody. Sir George had lost a hundred and thirty
napoleons to him the night before, and says, if he went
on, he should lose all his pay. Buonaparte had ingratiated
himself with everyone in the course of the voyage, and was
universally popular.

* His brother Leonard's house.

Sir Hudson Lowe is to take out for him to St. Helena a considerable collection of books, many of his own particular choice, especially mathematical works, and a set of the best French translations of the classics. I hope I shall hear from you soon, and from Mary too.

Ever affectionately yours,

FRA. HORNER.

————

To Dr. Marcet.

White House, Edinburgh, November 10th, 1815.

MY DEAR SIR,—I had not heard of your return to England until Mrs. Horner received Mrs. Marcet's very kind letter a short time ago. We were very much pleased to find that your journey had afforded you such high gratification, and indeed you were most fortunate in setting out at the time you did as events turned out, though it shewed you and Mrs. Marcet especially, to have very adventurous spirits. By this journey, and that to Paris last year, you have enjoyed the most interesting periods of the last two eventful years. We were sorry to learn that you found so much distress at home on your return. . . .

It is hardly possible to think of anything else but the extraordinary state of France. Jeffrey has just returned from Paris, and he gives a dismal account of it indeed. He had an opportunity of seeing leading men of the most opposite parties, and his impression is that a civil war cannot be far distant, and that the Duke of Orleans is very likely to be called to the throne. The conclusion I have come to, from the narrow sources of information now within my reach, is that the present King ought not to be permitted to reign, and I think it would be of the first importance to the cause of liberty all over the world if it were shewn that a

hereditary monarch has no claim to the crown, when the
people for whose benefit the office is created, consider it for
their advantage to set him aside.

I am now leading a very quiet retired life, in some degree
new to me, but affording many enjoyments. I look back
with much regret to the opportunities I had in London
of enjoying the society of many valuable persons whom my
little scientific pursuits brought me acquainted with, nor
am I likely to have their place supplied here. The Geological Society has now met, and I hope with a favourable
prospect of a successful Session. I am afraid that there
is some danger of its declining, or of its getting into bad
hands from the want of a few active persons who can
devote their attention to it . . . .

I shall be glad to hear what you are engaged in. I
hope you are not idle. And if you will now and then
tell me what is going on, it will be a great piece of charity.
In the report of the proceedings at the laying of the
foundation stone of the buildings for the London Institution,
I observe a Mr. Butler states that Davy has discovered
a mode of preventing the fire damp in mines. This would
be a discovery of vast importance, pray tell me what it
is. . .

Dugald Stewart has been for some days at a village
within a mile of Edinburgh, where he took lodgings for
the purpose of correcting the proof sheets of his History
of Moral Science in the Supplement to his "Encyclopædia
Britannica," now publishing. It is said to be the best
thing he has ever written. Playfair is to write the "History
of the Physical Sciences." Davy was applied to, to write
a distinct article on the History of Chemical Science, but,
I understand, declined it. Pray remember me very kindly
to Mr. William Haldimand, I hope he has got quite
well. Mrs. Horner sends her kind regards to yourself

and Mrs. Marcet. Mary is very well, and sends her love
to Louisa and Sophia.

Believe me, my dear Sir,

Yours most faithfully,

LEONARD HORNER.

———

From Dr. Marcet.

London, *November* 18*th*, 1815.

MY DEAR SIR,—I am glad Mrs. Marcet has told Mrs.
Horner something of our German expedition, for I should
not dare to touch upon that subject, which always draws
me into interminable stories. We returned convinced that
Switzerland was still the happiest spot under the sun,
though England is the greatest and the best of all countries.
As to France, we never entertained very strong hopes of the
King preserving his throne, though we wished it sincerely for
the sake of the peace and repose of the world. As to the
principle of divine right, I hold it as absurd as you do,
though I thought upon the whole, its operation in the case of
Louis the Eighteenth rather desirable. These poor French
are almost as poor in fit persons to govern, as the Geological
Society is in presidents. The Duc d'Orleans is generally
admitted to be a fit person, but then, the very act of
snatching the crown off the head of his relation, to whom he
has twenty times sworn fidelity and allegiance, would make
him a traitor, and therefore a man unworthy of the throne.
And if the King were ever induced to resign the crown in his
favour, would not the bigots of the South and the fanatics of
the West, rise in favour of his nephew? Wherever we turn
our looks therefore, the affairs of France appear desperate,
and if it were not for the English bayonets, I have no doubt

that new convulsions would immediately take place, yet as
there is unquestionably a spirit of liberty strongly implanted
in some French heads, and as the nation, with the exception
of the late army, is certainly for the present cured of the
thirst for war, it is possible that the cause of liberty might
ultimately gain by leaving the French, after all this bleeding
and sweating, once more to themselves.

Davy's contrivance to avoid explosion in the mines,
is shortly this :—It has long been known that the gas
in question does not take fire unless the proportion of
air mixed with it be considerable, or in other words,
that the quantity of oxygen present, must bear a certain
proportion to that of the gas, in order that combustion
may take place. Sir Humphry has taken pains to ascertain
accurately the limits of this susceptibility of ignition, and
having done this, has endeavoured to introduce all round the
candle or lamp which is used in the mines, an atmosphere so
vitiated, that gas coming into contact with the light burning
in it, should not be explodable. Now, such a vitiated atmosphere obtains in a lanthorn in which the access of air is so
scanty, that the light within it burns rather dim, though it
does not go out.

Davy's contrivance, therefore, simply consists in using a
lanthorn (of the most common possible construction), having
such a small aperture both at the top and the bottom, as
shall make the candle within it burn at its minimum of
brightness. I have seen the leading experiments performed
by Davy, and was perfectly satisfied that a current of the
gas in question might be introduced with safety into that
lanthorn, since the candle was almost immediately extinguished by it, without any explosion or combustion of the gas
taking place. Davy's expectation, therefore, is that the going
out of the candles will not only prevent all explosions, but
will afford to the miners a valuable test that they must retire.

Pray remember me most kindly to Mrs. Horner, to your sisters, and all your family. I will not talk any more of paying a visit to Scotland, but I will do it the moment it is in my power.

Mrs. Marcet begs to be most particularly remembered.

Yours ever most truly,

A. MARCET.

To Dr. Marcet.

Edinburgh, 21*st November*, 1815.

MY DEAR SIR,—I went last night to the Royal Society to hear Murray's paper read describing his plan for preventing the inflammation of the fire damp in coal mines, and as there has been a remarkable coincidence in Davy's bringing forward a similar paper, you will be anxious, I daresay, to know what Murray's plan is, and I shall give you as exact an account of the paper as I can. After stating that the most improved system of ventilation has been found insufficient, and that any attempt to destroy the gas by chemical means is very unlikely to succeed, he says that the means most likely to be effectual in preventing accident, is to 'adopt some mode of lighting the mines, by which the gas shall not come in contact with flame. He also suggests the possibility of lighting the mines by means of a jet of coal gas being introduced into fixed lamps on this construction, that there would be so vast a supply of light at a moderate expense, as to render it unnecessary to carry lights about the mine. There would be every facility of constructing a gas apparatus in the mines, and they could not only use the small coal for this purpose, but by converting it into coke, render that valuable, which is not only useless at present, but removed at great expense. Such is the outline of his plan, which certainly is very simple, and in theory admirably calculated for the purpose; whether it will be found to answer when submitted to experiment remains yet to be proved. Brande has sent to Mr. Allan a sketch of Davy's plan, which has some resemblance to that of Murray, but I think the latter has many advantages, judging from Brande's report.

Mrs. Siddons is here just now, playing for the benefit of her grandchildren, She is a little changed in appearance, but is as great in her acting as ever.

Mrs. Horner unites with me in kindest regards to yourself and Mrs. Marcet.

I am ever, my dear Sir,

Very faithfully yours,

LEONARD HORNER.

From Dr. Marcet.

London, *January 20th*, 1816.

MY DEAR SIR,—I have to congratulate you upon the birth of another daughter,* which I do with all my heart. Pray remember us both most kindly to Mrs. Horner. Having thus settled the business of your youngest child, you will naturally expect me to give you some news of another child of yours,† whom you left rather in jeopardy in Lincoln's Inn Fields, after having half completed his education. I am sorry to have no very favourable accounts to give you of him; for though he looks rather blooming, yet I do not rely much upon the vigour of his constitution; I doubt (to use a sporting phrase) whether there is *bottom* enough in him to proceed safely in the world, and I observe a sort of fickleness in his conduct which makes me doubt

---

* His third daughter, Susan   † The Geological Society.

whether you have not abandoned him to himself too soon, especially when I consider the vicissitudes and difficulties to which its infancy was exposed.

As, however, I am but a poor hand at keeping up a long metaphor, I shall now tell you in plain language, what has been done by the Geological Council. We were as usual approaching the critical period without being at all prepared for it. Warburton decidedly going out of office, Blake declaring his incapability of finding any person to replace him, your friend MacCulloch longing to sit in the arm-chair, and having a man ready (Mr. Stokes) to act under him as secretary, in case he (MacCulloch) should be elected president; on the other hand, Blake declaring his wish to retire, but at the same time his readiness to continue in office if the Council should think it proper to require his services for another year. And lastly, some members, such as Greenough, and Warburton, strongly inclining to let Blake retire, and even rather recommending it; such was the state of the Council when we met this morning. I ought to add that the expedient you had recommended, that Greenough should be requested to act as secretary under Blake, had been tried in vain. The result of all this was that MacCulloch was recommended as president, with his friend as secretary, and Blake was suffered to retire, not, I confess, without my regretting him exceedingly, for he filled his situation with every possible attention to the duties of his office, and every qualification to promote the objects of the society. There seems, however, to be but one voice respecting the claim of Dr. MacCulloch to the chair, and I am one of those who think most highly of his talent, but I thought he might have waited another year, without the society suffering from the delay, since the chair was so well occupied.

Pray if you find an opportunity, thank Mrs. Stewart for

her kind note to me, and Mr. Stewart for his intended
present to Mr. Prevost of a copy of his Introduction, which
is considered here as one of his most valuable productions.

I have already begun to peruse it, but I confess I am
but a slow, and unworthy reader, though a most zealous
admirer of all that comes from that profound and engaging
writer.

Believe me, my dear Sir,
Most sincerely yours,
ALEX. MARCET.

From the Same.

London, *March 7th,* 1816.

MY DEAR SIR,—The Medical Chirurgical Society, is going
to publish an account of the disease of which Dr. Saussure
died. It was a palsy, attended with some peculiar circumstances. The case is drawn up by Dr. Odier. When I was
in Edinburgh (alas, some twenty years ago) I was asked by
M. de Saussure's family to lay his case before Dr. Black,
and to request his opinion upon it. Dr. Black expressed
himself as strongly interested, in the welfare of the great
naturalist of the Alps, and desired me to call again upon him
in a few days, when he put into my hands as an answer to
the questions respecting de Saussure's case, the history of
of an attack of palsy which his friend, Mr. Ferguson,
experienced about fifty years ago, and of which he has been
for nearly the whole of that period perfectly recovered. This
history of Mr. Ferguson's illness is drawn up with Dr.
Black's peculiar conciseness and simplicity, and both from
the importance of the subject, and the peculiar interest
which either the narrative, or the celebrated literary
character, who is the object of it, possesses, it appears
desirable that this case should be thrown into a note

into Odier's account, to which it essentially belongs. Now, I would, without scruple, print the document in question, provided Mr. Ferguson's name did not appear; but, as I conceive, the name would add considerable interest to the narrative, I write to you to know whether you think I may let Mr. Ferguson's name appear without its being considered by him as an undue liberty, a reproach which I should be exceedingly sorry to deserve.

We are going on pretty well with our new President and Council in Lincoln's Inn Fields. The meetings are rather more numerously attended, though I think by no means improved; the change of officers cannot fail to occasion some change in the set of men who usually attend, and as all those with whom we were the most intimate were personally attached to our two last Presidents, we could gain nothing in that respect. There is, however, but one voice as to the claims and perfect competence of our new President. The new secretary reads very well, and seems very zealous.

Davy's last lanthorn is excellent. It is really a very pretty thing, both from its extreme simplicity, and from the very unexpected fact upon which its efficacy is founded. There is, I think, no doubt of its answering the purpose perfectly well.

We are here all doubt and suspense, about the fate of the Property Tax. I have very great hopes of its being either rejected or at least carried by such a small minority, as would be equivalent to a rejection. Your brother performs wonders, he becomes every day more dreaded and esteemed by those whom he opposes; and in his own party he becomes every day more conspicuous. It is a glorious thing to have such a brother; I understand the opposition were not displeased with the strength of their minority last night on the Army Estimates.

Louisa is now plunged deep into the arts of dancing, playing, and drawing; Frank fags hard at school, they all, with Mrs. Marcet at their head, beg to be remembered to Mary, and with our kindest compliments to Mrs. Horner,

Believe me ever, my dear Sir,

Most sincerely yours,

A. MARCET.

To Dr. Marcet.

Edinburgh, *March 14th*, 1816.

MY DEAR SIR,—My medical friends here speak in the highest terms of the usefulness of the Medical and Chirurgical Society, and as you were, I believe, one of those most immediately concerned in the establishment of it, such praise must be as gratifying to you, as it is to me when bestowed on its *fellow lodger*. I wish, however, that the Geological Society would communicate their papers to the public with the same despatch that the Medical Society does, and I see no reason why it should not. There is some unaccountable negligence in this respect, and it greatly diminishes its usefulness. Although MacCulloch may not be so pleasant at the Council Board as Blake, I hope you will not fail to attend as often as you can. You are one of those valuable guardians of liberty, that must not be allowed to slumber, and for God's sake do not allow the Council of the Geological Society to degenerate. If your present Council do their duty there is nothing to fear. Aikin was so kind as to send me a list of the new elected officers; when I read it to Playfair he said they ought to do a great deal, for it was a Council fit to govern the world.

Major Stewart (or, I believe, more correctly, now Lieut.-Col.!) has only been three hours in Edinburgh since he came to Scotland, and I have not seen him. I believe he left London sooner than he intended, in consequence of an

accident Mrs. Stewart met with ; she fell from a chair, and
either broke, or severely dislocated her arm. We had no
idea that your little boy had continued so long ill as appears
from your letter, and still more from one that Mrs. Horner
received a few days ago from Mrs. Schmidtmeyer. I trust,
however, that there is now no doubt of his recovery, and
that you will be able to restore him to perfect health.

How greatly Davy rises when he chooses to exert himself.
He has been spurred to this last effort of his ingenuity by
Murray and Clanny, and he appears to have outstripped all
his competitors. It will be found successful in the mines, I
trust.   Mrs. Schmidtmeyer's letter was a most entertaining
budget of news concerning those we care most about, and she
treated us with the account Lady Davy gave at your house
of her Knight's exhibition with Biot. I have heard a story
of De Candolle having been turned out of his Professor's chair
at Montpellier, by the enlightened and liberal Louis le Desiré,
and that his visit to London was occasioned by that circumstance. I hope his return to France is a favourable
symptom, as the case, as I heard it related, was a very
grievous oppression.

We wait, with some anxiety, for the arrival of the mail
to-morrow, which will bring, we expect, the result of the first
struggle on the Property Tax. If ministers are beat it will be
a glorious triumph to the country, more valuable than a
thousand times the *sum* contended against ; nor will the good
effects be confined to this country—the cause of the people
will gain by it in other countries of Europe. But as to the
losing of the question shaking ministers in their seats, as the
Morning Chronicle talks of, it is a very hopeless case. They
have long lost the shame of being in a minority, and the
Regent cannot do without them, for who would he find, with
better principles than they have, to submit to his profligate
expense, and go through all his jobs.

You are very kind in what you say about my brother, who appears to have made considerable progress in the House of Commons this Session. It is the highest praise that can be bestowed on him, that he commands the respect of those whom he opposes, and I think he may safely be held up as an example that a Scotchman may be pure in politics.

You have probably heard of Dr. Somerville's appointment to the Army Medical Board. He is a very good fellow, and his wife is a very interesting woman. She is a person of very extraordinary acquirements, particularly in mathematics. But she has not a shade of blue in her stockings. They will be a great loss in the society of Edinburgh, and Mrs. Horner and I have hardly yet formed her acquaintance. They have taken a house in Queen Square, and set out for London to-morrow. She is desirous of being acquainted with Mrs. Marcet, and my sisters have requested me to ask the favour of Mrs. Marcet to call upon her when she gets to town, which I do with pleasure, as she is one that excites great interest among all who know her. She will send her card, with a note from my sister, to Mrs. Marcet, when she reaches London. We are beginning to enjoy our visit to London by anticipation. I hope we shall be there early in June. Mrs. Horner unites in kindest regards to yourself and Mr. Haldimands, sen. and jun.

Ever most faithfully yours,

L. H.

From Francis Horner.

Taunton, [*March 30th*, 1816.

My Dear Leonard,—The enclosed letter to you from Mr. Poole has been in my hands for some days after having made a journey to town. I detained it, by his permission, to read it, for which purpose he left it open. The distinction he has pointed out between the Saving Banks and Friendly Societies, is just, and important to be kept in view. His reasoning about the poor laws is not so sound or distinct, though the practical conclusion, in favour of retaining those laws in England, is the opinion which I have long held, but upon principles somewhat different; in one respect, indeed, widely so, for whereas he seems to approve of the modern conversion of the poor rate into what he calls a labourers' rāte, I look upon that recent change in the system to have been one of the most pernicious corruptions it has ever undergone, and to be upon all principles, both moral and economical, quite indefensible.

I look forward with much pleasure to the month of June. My kindest remembrances to Anne and Mary; my acquaintance with Fanny will date from my next sight of her. I hear much of her beauty and good humour.

Ever affectionately yours,

FRA. HORNER.

———

From Dr. Marcet.

April 15th, 1816.

MY DEAR SIR,—As I expect to have the pleasure of seeing you in London very soon, I should not have troubled you with a letter at present, were I not tempted by your kind offer respecting the late Mr. Ferguson, to beg of you to collect for me the particulars of his death, and of the state of his head, if it has been examined.

Your last letter was full, as usual, of sundry matters of great interest, and I can assure you that if I were not under a constant state of feverishness, occasioned by want of time, and a perpetual remorse of leaving a number of things undone, I would be writing to you incessantly, in order to receive your answers. I hope you illuminated your house, or at least you gave free vent to your joy, in some ostensible shape or other, upon the occasion of the grand and ever-memorable victory. Of Waterloo? No—a much greater victory than that, I mean the extermination of the Income Tax, that is, of the greatest enemy to political liberty, to independence of character, to old English feeling, and to lasting peace, that ever existed at any period of the world. As for *us* philosophers, chemists, and scientific amateurs, we celebrated this event by generous libations of champagne at the Chemical Club. The two members who showed the most spirit upon that occasion were Dr. Wollaston (the Pope!) and myself. Being rather out of practice, we both exceeded a little the capacity of our brains, and as we rose from table, we discovered by certain vulgar symptoms, the fragility of our nerves. This excellent wine was the produce of a wager, which I won against Blake, who had bet two to one that the infernal Tax would pass, though he detested it nearly as much as myself.

The Geological Society sends you its kindest regards; a good house is in contemplation in Bedford Square, Covent Garden, a situation which I do not much admire.

Adieu, my dear Sir, come as fast as you can, and believe me, ever most sincerely yours,

ALEX. MARCET.

P.S.—Mrs. Marcet has seen Mrs. Somerville, and likes her very much; she will be a great acquisition to our neighbourhood.

To Dr. Marcet.

Edinburgh, *April 24th*, 1816.

MY DEAR SIR,—I am happy to say that the time begins to draw near when we shall set out for London. I trust we shall be in London by the first week of June. The party will consist of Mrs. Horner, Mary and myself—our round, jolly Bacchus, Fanny, we grieve to leave behind, because she is at present our daily amusement, and we shall miss her very much, and there is a little disappointment to our vanity, for we think her very magnificent. There is only one difficulty, the same which prevents Lord Castlereagh from realising all his great schemes. As to the Bambinella, she is not yet in that state of maturity as to entitle her to be talked of out of the nursery. I am anxious to get to town to see my brother for some time before he goes on the Circuit, and to see something of London before it is empty. I hope you will not go very soon to Clapham, it will be a great disappointment to Mrs. Horner, if she only gets an occasional glimpse of Mrs. Marcet, I hope the backwardness of the spring will in the same proportion retard the period of your migration. I shall be delighted to see the bright original of *the* picture, for she must be a very interesting person.

You will be much pleased with the article in the *Edinburgh Review* just published on Davy's lamp; it is by Playfair, and in his best style. It is a high reward to Davy of itself. No man has ever deserved a proud distinction with a juster claim; there appears to be the strongest reason to hope that he has destroyed the formidable enemy, and he ought to participate in the wealth which he has saved from a state in which it seemed utterly lost.

We are much obliged to Mrs. Marcet for her kindness in calling so soon on Mrs. Somerville. Since they went to London his appointment has been taken from him, we hear; one of the instances in which ministers have redeemed the Regent's pledge of economy—very justly too, I think—it will not be long before they must come to the reduction of some of the large items of their profligate expenditure. I put my shoulder to the wheel—I was one of the infinitely small quantities who composed the thundering power that knocked

down. Vansittart's Income Tax, for I attended at a tavern for
five days to receive signatures to a petition. It is the
greatest triumph the country has obtained over the Court
for a very long time, and will secure peace to us more
firmly than a thousand Congresses and ten thousand
treaties. . . . . You will see Mary much
grown and stouter, I hope; she will be sorry if Louisa
has forgotten her. Give our kindest regards to Mrs. Marcet.

Yours very faithfully,

LEONARD HORNER.

[In June of this year Mr. and Mrs. L. Horner, and their little
Mary, went to London, and at first paid a visit to Francis
Horner at his house in Great Russell Street. The state of
his health had begun to cause his family much anxiety.
Towards the end of the summer he joined them in Scotland,
and he, with Leonard and his wife and their three little girls,
went to stay at Dryden, a pretty place near the Pentland
Hills, about seven miles from Edinburgh, which his father
had taken for the autumn, that he might have quiet and
rest with his family, and visits from a few intimate friends
in Edinburgh. Francis Horner's illness was thought so
serious, that the doctors he consulted positively forbade him
from pursuing his profession during the winter, and advised
him to spend the cold months of that season, and the
following spring, in a warmer climate. His brother Leonard,
with the consent of his unselfish wife, offered to accompany
him wherever it was settled he was to go, and it was finally
decided that he should pass the winter at Pisa. Accompanied
by his brother Leonard he left Dryden on the 5th October,
and accompanied by his intimate friend John Murray, they
travelled leisurely south.]

To Dr. Marcet.

Edinburgh, *October 1st,* 1816.

MY DEAR SIR,—We got home in eight days from the time
I saw you, and after being there a few days, we joined my
father's family at Dryden, where we remained till Saturday
last. Dryden is a very handsome place near Roslin,
belonging to Sir Charles Lockhart (a minor), the house of
which my father hired for two months, for the sake of my
brother's health, that he might have the advantage of good
air and quietness, without being exposed to the fatigue of
Edinburgh society. He came there from the Circuit, not
much improved in appearance from what he was when I
parted with him in London, his cough still troublesome, and
his difficulty of breathing greater than before. Unfortunately
we have not had two days together of fine weather since he
came, which has had visible bad effects in increasing his
breathlessness; but the quietness and regularity of living
have diminished his cough a good deal. I make those
inferences from observing that he always breathed more
freely when the air happened to be warm, and excitement
generally brings on a fit of coughing. For a considerable
time he refused to have any medical advice, but he yielded
to our urgent requests, and sent for Dr. John Thomson and
Dr. Gordon, both of whom he had seen before as friends.
We had every reason to place great confidence in their skill,
and we knew that from their personal attachment to him,
they would bestow more than ordinary pains upon his case.
Their apprehensions at first were that he had water in the
chest, but after a very careful observation of his symptoms,
they hesitated to ascribe his breathlessness to that cause,
and frankly owned that there was a great deal of ambiguity
in the case. There is no doubt about there being a great
derangement in the digestive organs, but they do not

consider that that will account for his most alarming symptom, they fear that the lungs are affected.

Notwithstanding this state of his health, my brother determined to go to.the Somerset Quarter Sessions, and had fixed his day of departure. When Thomson saw this, he considered it necessary to tell him the extent of his danger, and that he requested for his own sake, and for the sake of those around him, that before leaving Edinburgh, he would take the best medical opinion it could afford, by having a consultation with Gregory and Hamilton. He consented to this, and last Sunday they gave their final opinion to this effect; that there is every appearance of such an affection of the lungs, as to render it absolutely necessary that he should abstain from all business or exertion of any sort during the next winter, and that he ought, without hesitation, to seek the benefit of a warm climate, that as yet no mischief has been done, but that unless such precautions are used, there is no doubt that the most serious consequences must ensue.

This decision was a considerable blow to him, as you may easily conceive such an interruption to all his views would occasion, but he has determined to adopt their recommendation in its fullest extent, and to go abroad without delay. As the fatigue and anxiety of such a journey must be lessened to him as much as possible, and as it is important that he should have some one near him, to watch carefully over him, I have determined to accompany him. Fortunately I can do so without injury to my business, and although the separation from my family creates many uneasy feelings on both sides, all such considerations of that nature, of course give way. We propose leaving this on the fifth for London, where the ultimate place of destination will be fixed. My brother talks of Naples as most likely to offer the greatest advantages, and he thinks that by crossing Mont Cenis we may be at Turin before the end of this month. I shall take the liberty of

applying to Mr. W. Haldimand for his assistance in making
my arrangements for the journey, and I hope he will not
consider that by so doing I shall encroach too much upon
his friendship. I shall be much obliged to you, my good
friend, if you will turn your thoughts towards collecting any
information that you think may be useful to me in the
journey, for as I have not travelled in France, I shall have
need of a great many instructions. It is the farthest thing
from our intention to travel as Englishmen with our purses
open, but to go as economically to work as the necessary
comforts for an invalid will admit of. Our object will be to
get within the warm atmosphere of Italy as soon as possible,
and the road we shall probably take will be Paris, Lyons,
Chambery, and over Mont Cenis to Turin. I shall also put
my friend Prévost under requisition. We shall be at Holland
House during our stay in London of a few days, where we
shall, no doubt, get a great deal of information that will
be useful.

I have read the " Conversations on Political Economy "
with very great satisfaction. Mrs. Marcet has executed a
difficult task with great skill, and has made the subject not
only very accessible to ordinary readers, but at the same
time very interesting. She has discovered a royal road to
political economy, and I am sure it is a book which will
do an infinite deal of good, by removing many popular
prejudices, which are very deep rooted. Mr. W. Haldimand
cannot do a greater service to his country than by getting
each of his brother directors to study this treatise; and if
Prévost can effect the same purpose at Lloyd's, it will change
the whole character of the merchants of London, which I
suppose never stood more in need of improvement.

Mrs. Horner desires me to say, that with a very few
exceptions, it is quite level to the capacities of the ladies,
and that even these passages only require a little borrowed

light, to appear quite distinct. She unites with me in kindest regards to Mrs. Marcet, to yourself, and all your family circle.

Your's, my dear Sir, very faithfully.

LEONARD HORNER.

From His Mother.

Dryden, *October 4th*, 1816.

MY DEAR LEONARD,—As I may not have a proper opportunity of expressing my feeling on the present trying occasion, I trust what I say in this way will have the same effect. Yours and Anne's kindness we never can forget; indeed she has given a proof of affection towards our dear Frank that very, *very* few wives would have done, in parting with a husband she tenderly loves, for so long a time. I humbly hope that she will be rewarded by the comfort she and *you* have afforded to the rest of the family; because, however painful it is to part with my dear Frank, it would have been ten times more severe had he gone alone—or with any one we could not depend on—no one is so fit as yourself, and I trust that he will be able to nurse you, should you be unwell.

I trust in a kind and merciful Providence, who has blessed me with two good sons; may neither of you ever forget that the Almighty alone can protect you, go where you will. I am no bigot, nor am I what the world calls a Methodist, but I have had the happiness of many pious instructors, and have witnessed the happy end of those who put their trust in the Most High. Don't, my dear Leonard, ever forget your duty to your Maker; for that affords a comfort, nothing in this world can bestow. Sunday, I believe, is too much a day of amusement abroad; be assured something more is required. God bless you, read this with that warmth of affection that I have written it, and take it as it is meant.

I hope that all will keep well in your absence, and that you and my dear Frank will *both* be restored to us in health and safety, and believe me, my dear Leonard, nothing will afford us so much pleasure as hearing from you when at a distance. God bless you both, and preserve you to your

Truly affectionate mother,

JOANNA HORNER.

From Francis Horner to his Mother.

Cornhill, *October 6th,* 1816.

MY DEAR MOTHER,—That you may not be without the letter to-morrow which you were promised, though you will get one to-day, I write this from the same place after an excellent night's rest, without a single cough since I left Dryden. The weather is most favourable. I have been ELEVEN hours in bed (neither of my companions has been much less), and we are now setting off, before 8 o'clock. Murray is going to survey the field of Flodden, and then to join us in the evening at Newcastle. Nothing can equal his kindness, and Leonard's about me. After I got to bed last night, I read your parting letter to Leonard, which delighted me, and filled my eyes to the brim. God bless you, my dearest mother.

Love to all,

FRA. HORNER.

From Mrs. Horner (senior) to Mr. Francis Horner.

Dryden, *October 6th,* 1816.

MY DEAREST FRANK,—The few lines from Cornhill written by my dear Leonard made us truly happy. God grant that you may continue to mend. Your cough is certainly better, but he says nothing about your breathing. I hope soon to be informed that you are relieved in that respect; the day favoured your departure, and I flatter myself that wherever you are, you may have fine weather, as we are not to judge exactly by *our* climate. Yesterday continued so fine that your father talked of remaining here, *to-day* it is dismal; no walking, no riding for poor Mary, but a thick fog; I wish Anne had brought little Fan with her to cheer us up. Your empty room quite overset me. Anne and Mary took possession of it, but leave us to-morrow with your sister Fanny, whose ankle is better. We shall be very anxious till we hear of you from London. I trust that every post, however, will bring us some information. Your poor father read Leonard's note six times over; and Mary keeps it in her bosom. I never knew how dear you are both to me, till now.

Much as I hope from a warm climate, eighteen hundred miles between us is a dreadful distance, and I conjure up a thousand fears I hope I have no reason for, but I trust much to the advice you have had, and also to some friends that you have a chance to meet with abroad, but above everything I trust to a kind and affectionate brother, who I know will study your comfort, and do everything in his power to render the plan beneficial. I hope that he will be careful also of his own health. So much of our happiness depend on you and him returning safe and sound to us, that I hope the love you have for us, will make both attend to my earnest request. May it please God to restore you, my beloved jewel, to our fond wishes, and that health and happiness may attend you and my dearest Leonard wherever you go. I hope you will not go more on the sea than what is absolutely necessary, if you should be ill on board of a ship you cannot leave it, and I dread your suffering in many ways. Once more I pray for the Blessing of God Almighty to attend you, and believe me your truly

Affectionate mother,
JOANNA HORNER.

To His Daughter.

Witham Common, *October 9th,* 1816.

My Dearest Mary,—I cannot allow your birthday to pass without telling you how rejoiced I am at its return, and my anxious hope that you may have a long and happy life. It is the first time, I believe, that I have been separated from you on this day, and I feel no small disappointment that I have not joined the merry meeting at Whitehouse, or contributed my share in the diversions. I shall never pass a day, while I am absent, without thinking of my very dear good child, and there are few things which would give me so much pleasure now, as to have you on my knee. Your uncle gave us to-day at dinner, a bottle of excellent claret to drink your health, which Mr. Murray and I did in a full bumper and your uncle in a tumbler of toast and water. Your uncle is very well, and bears his journey without fatigue, and without having caught any cold. I hope that he will very soon be restored to perfect health, for no man makes a better use of life than he does; he is in every way so excellent that you cannot think too highly of him, or love him too much. You know how very much he loves you. When you write to me tell me a great deal about our darlings, Fanny and Susan, and give them both a great many kisses for me. Give my kindest love to your dear mamma.

Mr. Murray has written to-night to your grandmamma; remember me to nurse. God bless you, my dearest Mary,

Your very affectionate father,

Leonard Horner.

———

From Francis Horner to Mrs. Leonard Horner.

Welwyn, *October 10th,* 1816.

My Dear Anne,—I received at Stamford this morning your kind note. You see how leisurely we have travelled, and we have made the journey most successfully in all respects. I certainly am at least as well now as when I left Dryden. I ought to say upon the whole, better; for I have had little coughing all the way, and have not any one day felt that oppressive langour, which sometimes comes upon me. I have not once known the sensation of fatigue. All this I owe to Leonard, who is the most considerate and ingenious, as well as kindest of all nurses. But I never mean to say anything about his kindness in this misfortune of mine, or yours, in proposing to let him accompany me, because I could never express in any degree what I feel.

Your note of directions about him, in case he should suffer at all, shall be put up along with another I have on the same subject from Fanny, and a larger epistle from Dr. Gordon. But I trust he will have no occasion for any discipline.

I hope you had a happy day yesterday. We expect to reach Russell Street soon after twelve to-morrow morning. Murray is gone there before us; I have appointed Dr. Warren at two, and as I wrote to him from Morpeth, and gave him all the particulars of my story, I rather hope he will give a definite opinion at once. I cannot doubt that the ultimate decision will be for sending me abroad, though there may be some difference of opinion.

God bless you; my love to the dear children,

Affectionately yours,

FRA. HORNER.

[They arrived in Great Russell Street on October 11th, and the next few days were spent in Holland House.]

Lord Holland to Mr. Horner's Father.

Holland House, October 24th, 1816.

DEAR SIR,—Your letters shall be duly forwarded to your son. All his friends here are sensible of the advantage of his kind and affectionate brother Leonard being with him, and cannot but admire the disinterestedness of Mrs. L. Horner in promoting such an arrangement.

We had the satisfaction of finding the opinions of the London physicians less unfavourable than those of Edinburgh had been represented to us; and though I am not sanguine enough to think there is no cause for uneasiness, I certainly parted with him with a better opinion of his prospects than I entertained from the reports I had heard. His countenance was better, and both he and his brother maintain that he has gained flesh, which, if well ascertained, is a very consolatory symptom indeed, and one that outweighs very many of a less favourable description. No thanks are due to us from him or his friends. I am quite sure that nothing would make Lady Holland or myself happier, than the power in any degree of promoting the recovery, or contributing to the comfort of one of the best friends, and best men I ever knew. You may depend upon it, my dear Sir, that there is no man in England who has more sincere friends, as there is certainly none who deserve them more; and there is none who has greater pride in reckoning himself one, than your

Obliged humble servant,

VASSALL HOLLAND.

CHAPTER V.

OCTOBER, 1816—1817.

Extracts from the Journal of Mr. Leonard Horner to His Wife.

Paris, *October 27th,* 1816.

We reached this without any accident. A M. Montrond, whom we met at Holland House, a friend of General Flahaut, (the Aide-de-Camp of Buonaparte) most kindly offered Frank his rooms in the Place Vendôme, during our stay in Paris. General Flahaut wrote to his mother, Madame de Souza, to have everything ready, and accordingly we found on our arrival at St. Denis, that she had been there, and had left a note to say that everything was prepared. We, in consequence of this note, drove to M. Montrond's house, and are splendidly lodged—there was a good fire and beef tea ready on our arrival, all which things Madame de Souza had prepared in consequence of her son's letter, desiring her to watch over Frank as over her own son. M. Montrond's servant is a most obliging, active person, and his coachman understands English, so that John * is well taken care of. Within half an hour of our arrival we got an excellent dinner from a traiteur hard by, and after dinner Madame de Souza called for a few minutes to see that everything was comfortable. It is impossible to meet with greater kindness. I went last night to see the Palais Royal—all the shops were lighted up, and a great many people were walking round the Piazza—all lounging and amusing themselves. I went to the garden of the Tuilleries this forenoon, and saw the King appear at the balcony and bow to the people assembled below, as he passed from mass. He was accom-

---

* Francis Horner's servant.

panied by his brother, the Count d'Artois, the Duke and
Duchess d'Angoulême, and the Duke and Duchess de Berry.
A few voices cried out, "Vive le Roi," not so many as
I expected. The Palace itself, and the gardens, are quite
magnificent, indeed the whole town far exceeds my expectations in point of splendour. What I have seen, you may
believe has been very cursorily, but enough to give me
a great desire to spend some time here, and perhaps when we
return, I may see it with a little more leisure. The day
has been remarkably fine, the thermometer between sixty
and seventy degrees. Frank and. I, acoompanied by Mr.
Gallois (a most amiable person, a great friend of Sir Samuel
Romilly and Mr. Dugald Stewart) drove in a fiacre along
the Boulevards. Everything was gay and active; jugglers,
Punch, puppet shows, tumblers, theatres, and all sorts of
amusements, very unlike the Sunday evenings in the land
you are now in; but without being disposed to moralize, I
doubt very much whether there is not more substantial
happiness in the calm, decent quiet of the day of rest in our
country, than in all this fever of dissipation. It is, however,
a very pleasant scene to see so many gay, good-humoured
faces. A few of the shops were shut, but the greater number
of shopkeepers seem to be following their usual occupations.
I have seen bricklayers, carpenters, shoemakers, &c., &c., at
work as in ordinary days. The Revolution has effectually
destroyed the superstition of the Catholic religion among a
great proportion of the people of France; but unhappily it
has at the same time destroyed that religious feeling, from
which so much comfort is derived to the great body of the
people, and from which such solid political benefits are
derived.

October 29*th*, 1816.

I went last night to the Theatre Francais to see Talma.
The French tragedy is a thing so totally different from every

other dramatic representation, that one must have seen it
frequently to understand either the language, or the meaning
of the gesticulations.　Nothing can be more foreign to real
life than it is, and in the second-rate actors it appears quite
ludicrous.　The piece was Hamlet, said to be imitated from
the English, but it is a very distant imitation indeed.
Talma's acting is certainly very powerful, and were I to see
him three or four times, I should perhaps be able to see his
great merits.　Mdlle. Duchesnois, the great actress of the
French stage, played the Queen with great effect, but I
could not get over her unspeakable ugliness.　It is a very
dismal theatre, and is not in the least to be compared with
the splendour and gaiety of those in London.　George Dandin
was the farce, but I did not stay, as I should have been too
long away from Frank.　I must see Mdlle. Mars on my
way back.

Asti, *November* 15*th*, 1816.

Turin is a very handsome town, the style of architecture
is good, and the effect would be magnificent if the material
were better, but the whole town is built of brick plastered
over.　The streets are in general wide, and the houses lofty,
and there are very few mean looking houses.　The great
square in which our hotel was, is very grand indeed; opposite
to our windows was the king's palace, and on the middle
of the Piazza, the palace of the Duke of Aosta, the front of
which is considered a very fine specimen of architecture.
One of the finest theatres in Italy is here, but it is only open
during the Carnival.　I went to the Opera, where there was
the first representation of a new Opera Buffa, but neither
the singing nor acting were superior to the second rate
performers of the London Opera House; the dancing was
very poor.　The theatre itself was plain, and the total
absence of light from every other part of the house except

the stage, has a dismal effect. It belongs to the Prince
Carignano, whose box is in the centre of the house; he was
there himself, he is a young man about eighteen, with
a very sulky countenance: his palace is an immense
building of red brick, left in a very unfinished state.
Buonaparte built a new bridge of five arches over the Po,
it is handsome, and from it there is a most magnificent view
of the Alps, forming a crescent of a complete half circle.
Monte Viso, which is above Briançon, towering to a great
height above all the rest of the range. The Po is here a
considerable river, but neither clear nor rapid; one cannot,
however, look on it for the first time without great interest.
When Emmanuel returned, he proposed taking down this
new bridge because it had been erected by Buonaparte, but
it being represented to him that it was very much wanted,
after a good deal of solicitation, he consented to allow it to
remain, but made a vow that he would never cross it. The
French had established a Botanic Garden at an old palace,
built by a daughter of Henry IV. of France; the King's
gardener, who had accompanied him to Sardinia, when he
got possession of this garden, pulled up all the plants that
had been put in by the French, and threw them into the Po,
cursing them as Jacobins.

The streets of Turin are covered with priests, they are
distinguished by wearing at this season a large black mantle,
and a low cocked hat, similar to what some of our more
rigid Bishops wear. They have come back in vast numbers
with the king, and have great influence at court, but they
have neither got back their possessions nor regained their
ascendancy over the people, who do not hesitate to laugh at
them when they attempt to play their old tricks. But their
dominion is complete in many instances, for the press is
not only wholly under their control, but certain books are
not permitted to enter his Sardinian Majesty's dominions.

Machiavelli's works and Alfieri's are among the number, and a bookseller told me, that a bale of books he was receiving from Florence was detained at the Custom House on account of its containing some of Machiavelli's works. I was not able to find in the bookseller's shop any publication relating to the change in the Government on one side or another. The Convents, which were all abolished by the French, are filling up again. I saw two fat friars of the order of San Tomaso, strong fellows about fifty, each six feet high, dressed in brown, with a rope round their waist, and a rosary hanging by it, their cowls half over their heads. Asti is an old, ugly town with few good houses, among the best is the palazzo of the celebrated Alfieri, who was born here. His sister, the Countess of Cumiana, now lives in it. I went over it, and saw only large, comfortless rooms, with brick floors, wretchedly furnished, and yet the Countess has lately fitted it up. There is a very good picture of Alfieri himself, which he presented to his sister with the following inscription upon it :—

> " Non che a te fida Suora, ai piu remoti,
> Figli de' figli tuoi, mia sola prole.
> Questo mio volto interpreti i miei voti."

Opposite to the house of Alfieri is a very large monastery, which they are busy just now repairing. The cathedral is a large and rather handsome brick building, and the interior is covered with rich fresco paintings, some of which seemed to me to possess considerable merit. The wine of Asti is the most famous of that made in Piedmont.

Novi, 16th November, 1816.

Genoa, 18th Nov., 1816.—Novi is a poor place without any object of interest, and a stall with old Bibles and prayer books, is the best bookseller's shop. I went there to endeavour to find some description of the place, but without success. In the shop was a monk of the Order of St. Francis,

there being a monastery here. He was a young fellow, very
dirty, and poisoned me with an odour of garlic. During
the French Government these monks assumed the dress of
priests, but since the restoration of the Pope's power, they
have taken to their regular habits. Novi was the scene of
the great victory gained by Suwarrow over the French, the
battle was fought on the hills above the town. From Novi
to Genoa is the worst road we have passed over in our whole
journey, it is about forty-five miles over a mountainous
country the whole way, and the whole of the road is paved
and in the very worst·possible state of repair; we expected
every minute to break down, but we arrived without any
accident. From Novi to Voltaggio, which they make four
postes, notwithstanding a constant succession of hills and
the· miserable road, we went in less than three hours.
In summer this must be a beautiful drive, for there is a
constant change of scene, among hills covered with wood,
and full of deep ravines. From Voltaggio to Campomarone
we crossed the Apennines, through the passage called La
Bocchetta, celebrated for the magnificence of the view from
the summit, and equally notorious formerly for the robberies
that were committed upon travellers. For a long time,
nothing of the sort has occurred, and indeed the stories that
have reached England lately, are full of exaggeration. The
road over the Bocchetta is carried along the side of a stream
to within a very short way of the summit, but the descent to
Genoa is very rapid. very long, and the road so execrable
that we were'shaken to pieces.

The view from the summit is very grand, and we had the
finest weather to see it in. The Mediterranean, which I saw
for the first time, shone like a mirror; the Apennines present
the most varied and picturesque shapes; and the valley of
the Polcevera is covered with vineyards and olive trees, and
studded with the country houses of the rich Genoese. From

Campomarone there is a most excellent road, made by a
Doge of Genoa, very level, and a great contrast to that we
had just passed over. For a mile and a half you pass
through a suburb, but one of palaces, until you come to the
Lantern, a very lofty, light house, when you suddenly turn
the corner of a high rock, and Genoa bursts upon you at
once, in all its grandeur. It is the grandest thing I have
ever seen—a bay of above six miles' extent, lined with very
fine houses towering one above another, and the high hills
that form the screen behind the town, studded with those
white palaces, among the trees and vineyards. The blue sea
and the shipping are the foreground of this great picture.
Frank had written to Lord Minto (who has been here for
some time), when we determined to take this road, and just
as we entered the town, we met Lady Anna Maria with her
sister; they were walking with Mr. Hartup, a son of Sir
Edwards; the meeting was fortunate, for they saved us from
going to a very bad inn, and Mr. Hartup very kindly gave
up his room at the Hotel de Londres to Frank, on account
of its being in the best part of the house, both for warmth
and in point of height, being more easily reached than the
other rooms; for in Genoa the principal rooms are upon
the fifth and sixth floors, and it was of great consequence to
get one lower down. That we are upon, is the third, and
Frank was carried up in a sedan chair, for the staircase is
at least twelve feet wide. The house was formerly the palace
of some nobleman, the rooms are very fine, and the hall
and staircase are all of marble. It looks upon the bay and
shipping, and last evening we had a sunset like the middle
of autumn. From the time we crossed Mont Cenis, until
we got on this side of the Apennines, the weather has been
very cold, and the night before last the thermometer at Novi
was 29°, last night it was not lower here than 42°, and they
consider this as very cold for the season.

Lady Anna Maria has been sitting with Frank all the
morning. She is a delightful person, and I cannot tell you
how great the pleasure is to see and hear an English lady,
after so long being away from them. Genoa is the cleanest
town I have ever seen—Brock, in Waterland, does not excel
it in that respect; and the women are so tidily dressed.
They have all a muslin scarf thrown over their head
and shoulders in a most graceful manner, of the most
beautiful whiteness, even the middling ranks wear this,
the lower orders having the Mezara (the name by which this
piece of dress is distinguished) of a gay flowered chintz.

When we left Turin, we had not determined whether we
should attempt the journey from Genoa by land or by sea,
as we waited for accurate information at Genoa. Buonaparte's
Government had begun a road along the coast, which was
to have been practicable for carriages the whole way to
Leghorn, but unfortunately it has been left unfinished, being
completed only as far as Nervi, and a little beyond, about
seven miles from Genoa, and for a short way between
Chiavari and Sestri; the rest of the way as far as Sarzana is
only fit for mules. At Sarzana the new road again begins.
This road is however, frequently travelled, and ladies and
invalids usually travel in a Portantina, which is exactly a
Sedan chair, and carried by men; the distance to be travelled
in this way is about sixty-five miles, and is usually
accomplished in three days. But the difficulty of this mode
of travelling, the danger of exposure to cold in the bad inns,
and the encouragement held out to us of the great facility of
a sea voyage to Leghorn, determined us to choose the
latter.

We were very fortunate in finding the Minto family at
Genoa; I did not know any of them before, except Lady
Anna Maria. Old Lady Minto was living with the three
daughters in the Palazzo Brignole. Lord Minto, with all his

family have a house a little way from the town, and Captain
George Elliot, with his wife and three young children, were
in the Hotel de Londres, where we were. They have been
travelling through the South of France. We had them all at
tea in our rooms one evening, and Lady Anna Maria and
Lord Minto came the next night. They find Genoa very
dull and very cold, and are very sorry they went there.
Old Lady M. and Captain Elliot have commissioned me
to look for a house for them here. The young Lady Minto
is handsome and good-natured. Nothing could be more
friendly than they all were, and Lord Minto took a great
deal of trouble in making the arrangements for our voyage.
We were not a little surprised to find the Miss Berrys in the
same hotel, and still more so to find their old father walking
about the town. It was an adventurous thing to drag an old
man of eighty-four across the Bocchetta in the month of
November, but they are laying in a store for the *Blue
Parties* next year, and we hear that Lydia White is in
full force at Florence. I had some difficulty to prevent
the fatigue of an excess of kindness, and deep interest
on the part of these ladies for Frank. Sir George Clerk
quitted Genoa the day of our arrival.

By the recommendation of our banker, Mr. De la Rue,
we took our passage in the English brig, ' Rolla,' Captain
Simms, lately arrived from London, and going to Leghorn.
The wind had been blowing from the north for many days, a
quarter at which, in this time of the year, it continues pretty
steadily, and so favourable that we were told we might safely
reckon on not being more than one night on board. With
some difficulty we got our carriage shipped, and on Tuesday,
19th, by twelve o'clock we had all our baggage on board,
when the wind gradually fell, and then turned round to the
worst point of the compass for us. The captain said it was
of no use to sail that night, so that I had to get Frank's

H

bed landed and put up again at the inn, and some other
parts of our baggage. Next morning, at six, we were
summoned to go on board, and by nine we left the harbour.
It was a beautiful morning, and we had a good breeze
directly in our favour, which soon carried us a few miles
from the Mole. The view of Genoa from the sea is wonderfully grand; the white houses covering a great bay, and
towering to a great height upon the hills which enclose it,
the great Moles stretching into the sea, the tall light houses
and the domes of the churches, form a mass of striking
objects not often to be seen united. The forms of the hills
that immediately surround Genoa are fine, and the snowy
Alps are seen beyond them in the distance. The favourable
breeze which made us enjoy this fine scene soon died away,
and from that time till Monday evening, when we anchored
in Leghorn roads, we had constantly unfavourable weather;
sometimes we had a strong contrary wind, with a very heavy
swell, which knocked the ship dreadfully about; at other
times, we had the same swell, with a calm which was worse,
and we had a great deal of rain, which fell with the
impetuosity of a waterspout. We had, besides ourselves, in
the cabin a Sir Philip Belson, a colonel in the army, a very
gentlemanlike man, with whom we have been very much
pleased, and who behaved with a kindness and want of
selfishness which was pleasant to find.

It was early on Monday morning that we saw the island of
Gorgona, which gave us hope of reaching the end of our
tedious voyage (for the whole distance is but 90 miles), but
we were kept beating off with a wind directly out of Leghorn.
Next morning we got into the harbour. We were all proposing to go ashore, when the captain asked me for our bill
of health, a document which we had to procure at Genoa to
certify that we had not the plague. I produced the paper I
got from the British Consul at Genoa, when, to my dismay,

and that of all present, it was discovered that I had not got the proper document, a mistake that arose from an ambiguous expression of the Consul. The captain went ashore with his papers, and returned with the answer of the *Ufficio di Sanita*, that the ship and all on board, for want of this paper, must undergo quarantine, until an answer could be received from Genoa. I immediately sent a message to our correspondent, M. Dussony, who, after a great deal of difficulty, upon our representation of the case, obtained permission for us to land the following morning. During this negotiation, I had to go in a boat to the Quarantine Office with a guard who never left me, and I had to speak across a space separated by two walls, and all my letters and notes were fumigated before being delivered. The person who fumigated them, took them from me by means of a long cane cleft at the end. The following morning between seven and eight, four men came on board to examine the vessel, who not only made no examination, but seemed to have no other object than to get money—half an hour afterwards came a grand barge rowed by a dozen men, with a flag flying, and containing the doctor, who was to decide whether or no we were infected. He stopped at a distance of twenty yards from the ship, and we all appeared on deck. He counted us, and one of the fellows with a red night-cap who came first on board, called out *Tutti in buona salute*, the doctor went off, and we were told that our quarantine was at an end, and that we might land when we pleased. Such a piece of mummery I have never witnessed, nor a more idle exercise of authority. As for the real precautions that were used, we might have had all the plagues of Egypt for aught they knew. The doctor was scarcely gone, when a boat came alongside with half a dozen fiddlers, who struck up a merry peal to cheer us upon our regaining our liberty, and to take advantage of the fulness of our joy, and openness of heart on

the occasion.   On the other side of the ship, were a man
playing a violoncello, a woman singing in parts with the man,
and a boy playing the violin to accompany them—besides
these, there were numberless women in boats, selling grapes,
apples, walnuts and wine.   There was something very
cheerful in all this, and I could neither shut my heart nor my
purse-strings against these welcome visitors, ...
Unfortunate and disagreeable as this voyage was, I believe
Frank would have suffered more had he gone by the road
along the coast.   . . .   We saw Lord Carnarvon at Genoa,
who is travelling with two daughters and two sons; they went
that road, and we have seen them since our arrival at Pisa;
Lord Carnarvon says, that the road is so steep in many places
that he is persuaded the Porteurs could not have carried
Frank in the Portantina over them, and the inns were so
bad, that they hardly afforded shelter, and they could get
nothing to eat but some eggs.   It was besides very cold.
This was not the case at sea, for the thermometer never fell
under 50, and in the middle of the day it was above 60 deg.
We had got everything ready to start, when a custom-house
officer arrived, and said that unless we had our baggage
examined and sealed, we should be stopped at the Gate going
to Pisa, and have our trunks opened there.   Leghorn is a free
port, and they are very strict in examining everything that
passes into the interior, so that Custom-house officers are
stationed at every exit.   It required no ordinary patience to
have to unload the carriage and transport all our trunks, etc.,
to the Custom House, to say nothing of the irksomeness of
having our clothes, etc., turned topsy-turvy, and after all, to
have to fee several rascals for this mitigated search.   All this
detained us three hours, and in two hours more we got to
Pisa, over a very bad road, most exceedingly rejoiced to find
ourselves at the end of our journey.   Lord Carnarvon had
been so good as to secure rooms for us at one of the inns,

which we have found very comfortable. All yesterday I was
occupied in searching for lodgings, which are very scarce, and
in general very ill-adapted for the wants of the English, as so
few of the rooms have fireplaces. There is besides only one
part of the town where it is advisable for a person in search
of warmth to live, the quay on the right bank of the Arno,
which faces the sun; for the difference of temperature is
wonderful. However, I have been fortunate enough to get
lodgings on this quay, and though small, and up two pair of
stairs, are such as I think will be comfortable. The sitting-
room and Frank's bedroom, front the sun. We have been
lucky also to get them at a moderate price, for they are in
general high. We go to them on Monday. We have been
fortunate to find several friends here, Mr. and Mrs. Achard,
Madame Constant, her two daughters, and Miss Achard, are
just arrived to spend the winter here—I called upon them on
our arrival, and found that they had been expecting us;
they are come on account of Madame Constant, not that she
is ill, but as a matter of precaution. They are exceedingly
kind, and they are such excellent people that we shall have
great comfort in their society. Many very kind enquiries
have been made after you, and my dear Mary. Three Miss
Allans (sisters of Lady Mackintosh), their sister Mrs. Drew,
and her two daughters, are also here for the winter. One of
the Miss Drews is a very nice girl. The weather is delightful,
we have constant sunshine, and bright blue sky without a
cloud. Our room in the inn has little sun, and we have a fire,
but I found Madame Achard sitting yesterday working at an
open window without a fire in the room. I shall keep a very
regular register of the weather, and transmit it from time to
time. I have seen nothing of the town yet, except the quay
along the Arno, which is very fine. There are not many
good houses—but the river here forms a crescent, which
produces a very striking effect at various points from the

variety of forms of the different buildings, and there are
three bridges, one at each extremity, the other in the centre
of the crescent. The quay on the left bank, from the total
absence of sunshine, is a very great contrast to the gaiety of
the right bank.

The Arno is here about as broad as the Thames at Staines,
flows slowly, and is very muddy—as much so, I think, as the
Parret at Bridgewater. There is a great deal of bustle in the
streets, and a constant noise, for the people here speak as
loud as if they were all deaf. There is a great appearance of
poverty and wretchedness—the streets swarm with beggars,
covered with filth and vermin—there are about fifty galley-
slaves here, who during the day are employed to sweep the
streets; they are chained two and two together, and each
pair is accompanied by a guard with a musket. The thieves
are clothed in red, the murderers in 'yellow; the shocking
associations one has of their crimes, the degraded situation of
men being chained together like dogs, and the constant
clanking of their heavy irons, create a most unpleasant
sensation.

———

From Mrs. Dugald Stewart to Mrs. L. Horner.

Kinneil House, 14*th Dec.*, 1816.

MY DEAR MADAM,—I found a letter here waiting me from
Lady Anna Maria. I now copy part, merely to show you
that the truth is told to others, exactly as to yourselves.
After mentioning the exertion of calling to the postilion,
which gave them a more alarming idea of his first appearance,
Lady Anna Maria says: "Next day I was with him almost
all day, and instead of the sad apprehensions of the day
before, I came home with almost as sanguine an opinion of
his case as he himself seems to have, and above all, a
conviction from his perfect calmness and composure both of

look and manner, that he really has no fever. Next day
(yesterday) he was to have gone, but the wind changed about
the middle of the day, and he is still here. With the change
of wind came rather a milder air, and Leonard allowed him
to go out. We then went to see some of the lions of Genoa.
He went up-hill in a Sedan, and on level ground and down-
hill walked so stoutly, that it did our hearts good. He was
carried up the hundred steps of this house, to see the
collection of pictures, and in the evening Gilbert and I were
favoured with Leonard's special permission to come and see
him, and we found him not at all the worse for the exertions
of the morning. His brother says he appears much better
within this week, and has not coughed at all. It certainly is
a most incomprehensible complaint ; whatever it may be it is
not consumption. Mr. Horner could not have a better or
more rational nurse than Leonard, and we submit to his
decrees with the same obedience as the patient. Luckily he
has not been severe, and the first day I was a little afraid he
had let us be there too much, but Mr. Horner was tired of
his solitude, and was glad to hear the English tongue spoken
by English women. He might have had as much of that as
he pleased by sending to the floor below, where the Miss
Berrys dwell. Unfortunately their voices are not musical,
and Leonard ran away in despair when they were visiting
Mr. Horner, *I* believe for fear he should be too strongly
tempted to take them by the shoulders and shew them out,
he says, to go and hunt for Gilbert or me, to come and help
him. They are certainly not formed for sick nurses, and
their's is killing kindness. Leonard means to *bar* them out
to-day." She then tells me of their plans, and does not
close her letters till they are sailed, and " George returned to
tell how well Mr. Horner was, and that fortunately the
north wind, which is the only fair one, secures a calm sea."
I have now, my dear Madam, given you, without changing a

letter, all the details. You will see the freedom with which
it is written, and though not meant for being seen, as it will
amuse your *own fireside*, I send it without scruple. The last
words of the letter are—" Since I have seen him, my hopes
are more sanguine than I can reasonably explain."

I hope you are all well. Best and kindest regards. Tell
Mary I am following her example, and reading the " Arabian
Nights." With sincere wishes for the happiness of you and
yours,

> I am, my dear Mrs. Horner,
>> Yours most truly,
>>> HELEN STEWART.

———

To Dr. Marcet.

Pisa, *December 7th,* 1816.

MY DEAR SIR,—You may, perhaps, have thought that I
had forgot the promise I made of writing to you, as you kindly
wished me to do, but it would be unjust, for until this moment
it has not been in my power, We arrived at this place a week
ago, and have fixed upon it as our winter quarters.

My brother's health, I am sorry to say, is in no respect
improved since he left London, his strength is not so great,
and he has lost flesh. I hope all this is to be ascribed to the
fatigue of the journey; he bore it very well until we got to
Genoa, at which place we embarked in an English brig for
Leghorn, and had a most unfortunately tedious passage of a
week, the whole of which time he was obliged to lie in bed, to
avoid sea-sickness, from which he suffered a good deal the
two first days. This confinement and the want of due nour-
ishment reduced him greatly; he has recovered strength since
we came here, but he has still a good deal to recover to
replace him in the same state he was at Genoa. For eighteen
days he was almost entirely free from coughing, though during

that time he was exposed to very cold air in the passage of the
Bocchetta, but within the last six days he has been occasionally
troubled with it again. But what makes me most uneasy is
that the difficulty of breathing is quite as great as ever. He
will, however, continue to follow up the plan for the three
months prescribed by Baillie and Warren, unless he receives
contrary directions, in consequence of the statements we have
sent of his progress by almost every post to Holland House
and Edinburgh. And as six weeks of that period will be
passed in perfect quiet, it ought to give the remedies a *fair*
trial, which, perhaps, may not have been the case during the
fatigue and irritation of the journey. He sleeps, in general,
well, his appetite is good, his pulse seldom above 80, generally
76, and only occasionally has he much languor.

I have been thus minute in the account of my brother's
health, as I know the kind interest you take in it, and it
would be a considerable satisfaction to me to know what your
opinion is respecting his case, whether you think the symptoms still indicate what you apprehended when I saw you, viz.,
adhesion. But I am not unmindful of the very scanty and
imperfect evidence I am asking you to judge from.

We left London with the intention of going to Pisa, as the
most advantageous place for passing the winter months; that
determination was taken upon a balance of very contradictory
evidence. When we got to Paris, we found Mr. Ord, the
member for Morpeth, lately returned from Italy, and who had
made climate a very particular study, as he had gone there
in search of it. He condemned Pisa in the strongest way, as
being exposed to cold winds, and much damp, and strongly urged
our going either to Rome or Naples, adding that Davy, who
had passed a winter at each place, spoke most in favour of
the climate at Rome. We left Paris with our mind very much
made up to go to Rome, but arrived at Turin, we got such an
unfavourable report of the dreadful rains and cold damp

vapours of the great capital, and such an encouraging report of
the mildness of Pisa, that we returned to our original scheme.

If such a book does not already exist, I think an account of
the climate of the various places in the south of Europe, and
in the tropical regions where invalids are sent to, founded
upon accurate meteorological observations, would be a very
valuable addition to a medical library. Such a compilation
could be made in the course of a few years, if no documents
now exist. I was quite surprised at the ignorance of the
medical people both in Edinburgh and London, as to this
point; both talked of Italy generally as if we could not go
wrong, whereas from all I can learn, there is much more
severe cold in Italy than in most parts of England; perhaps
it does not last so long, but still it is equally severe upon an
invalid, who may receive irreparable injury, even from a short
exposure.

I have begun to keep a register of the weather, but my
observations have been confined to the thermometer, for you
will be surprised when I tell you that I have not been able to
find a barometer in this place, and am now seeking an introduction to the Professor of Chemistry, in order to try to
remove my difficulties. Unfortunately I did not bring a
barometer from England, and I despair of getting one here.
I have three of Malacrida's thermometers, two of them with
registers for the extremes of heat and cold, and I have been
obliged to make corrections for all of them; pray give
Malacrida a good scold the next time you see him.

We have been exceedingly fortunate in finding here the
Achard family with Madame Constant. We heard from the
Minto family, whom we saw at Genoa, of their intention of
coming, and I found them out the night of our arrival at Pisa
—they will be most valuable neighbours. Since the above
was written Mrs. Achard very kindly called to read a letter
she had received from Mr. Mallet of the 22nd November—

the latest intelligence that we have received from England is
the 1st of November, for our letters are gone to Rome. . . .
I was happy to see Madame Constant looking so well, as I
feared to see her with a much less healthy appearance. She
is as handsome as I have ever known her. When she speaks
much she gets flushed and agitated, and she is therefore
forced to live very much alone; but by-and-bye when my
patient gets stronger, and she is better, I hope we shall often
meet. We have another family of friends in Mrs. Drewe, her
daughters, and her sisters the Miss Allens (sisters of Baugh
Allen,) there are some more English here, but none that we
know anything of. Florence is full, but the greater number
are flocking fast to Rome, for the cold at Florence is excessive.
My brother had a letter from Lord Lansdowne a few days
ago from Rome, he urges him in the strongest manner not
to come there, for he says the cold is much greater than is
usual in England at this season, and the road from Florence
almost impassable. They remain there till Christmas, and
then go to Naples. Dumont is there, and, Lord L. says,
buying marbles. The Achards have a story that he is in
love with a Principessa. Playfair got to Rome on the 18th
of November; he is in a state of the greatest enjoyment.

I have said nothing about our journey to this place,
because we passed in so hurried a way through the country,
and at so unfavourable a season, that I am not entitled to
hold any opinion. But such as it is, you shall have it.
From Calais to Moulins, the country is excessively ugly, with
the exception of a detached spot here and there; from thence,
and particularly round Lyons, there is a great deal to admire,
but I think, without permitting an undue partiality to mislead
me, there is no part of France that I traversed, which in the
best season, can be equal to Somersetshire, or Devonshire, or
Kent—for there are no trees thicker than a broomstick,
xcept in the Forest of Fontainbleau, and there is so little

pasture that there can be no verdure—a hop-ground is an
infinitely finer object than a vineyard. Where does *la belle
France* exist? All the provincial towns I passed through, are
wretched, without any sign of moderate affluence—very
much like the country towns of Scotland, such as Selkirk
and Hawick, Dunbar and Stirling. Except at Calais and
Abbeville, all the inns we were at were detestable, and such
as you meet in the Scotch towns I have named (I of course
except also Paris and Lyons), and the innkeepers extortioners.
If I did not make a bargain I was invariably cheated, and
when I made the best bargain I could, we paid as much as at
the best inns in England on the North Road. Then they are
furnished only for the reception of peasants, they do not
seem to be in the least aware of the wants of the people
above the lower ranks. I was very much struck with the
whole of Savoy. It was the first time I had ever seen
mountains on so grand a scale, and they have impressed me
with ideas of magnitude in the *Alpine School of Geology* that
I had no conception of before. The pass of La Chaille and
of Les Ecluses are wonderfully grand, and I was equally
interested in passing through the long, wild, valley of La
Maurienne. I longed very much to be travelling *marteau en
main*. We had a beautiful day to pass Mont Cenis, the sky
was quite clear, and the highest summits were seen. Greatly
as my admiration of Napoleon was raised by this magnificent
road, I am told that it will be increased ten-fold when I
return by the Simplon.

We went from Turin to Genoa by the Bocchetta, and had a
miserable voyage of a week to Leghorn. Nothing that I
have seen has been equal to Genoa. I cannot remember
ever to have been so struck as I was when the first view of
that splendid town burst at once upon us, on turning a hill.
I have only room now to send my very kind regards to Mrs.
Marcet, Mr. Haldimand, and Mr. W. H., to whom I wrote

from Genoa. I beg you will write to me very soon, and I hope you will send me a good account of your family.

Remember me very kindly to Roget, to Wollaston, and to MacMichael if he is still in London.

Believe me, my dear Marcet,

Yours with great regard,

LEONARD HORNER.

P.S.—My brother desires me to assure you of his very kind remembrances of yourselves, and Mrs. Marcet.

Journal resumed.

Pisa, *December* 28*th*, 1816.

This is a dull town, very little stir, no trade, and very little money circulating. It is wonderful to see the poverty of the shops of all kinds, but especially in all articles of luxury. There is not one jeweller's shop, there are a few small working silversmiths, who have a few show-cases of a coarse sort of finery, in silver and gold, for the bourgeoisie, and in large gilt necklaces, earrings and crosses for the peasantry. The booksellers' shops miserably provided, and no one shop where music is sold, or musical instruments, in Pisa. There is no such thing as a map to be bought, not even of Italy. There is a very neat theatre, which is sometimes used for the representation of operas and sometimes of plays, but I hear that it is not well attended. The town offers very few resources of amusement, either in itself or the country round, which is a dead flat for many miles, so that before you can mount high enough to see the objects around you, you must go at least four miles to the hills of San Juliano. A very fine range of mountains bound the view to the north. Their forms are quite Alpine, and they are grouped in the most picturesque manner. In front of these, towards the north-east, are a range of lower hills, also very beautiful in their shapes, and covered in many places to the top with plantations of olive trees. Nothing can be more

beautiful than the drive from Leghorn to Pisa, in regard to
the distant view of those mountains and hills; and the first
view of Pisa at the foot of the olive-clothed hills, as it
appears, is very striking; very like the view of Wells as you
approach it from Glastonbury. About two miles from Pisa
there is a farm belonging to the Grand Duke, of very
considerable extent. Part of it is employed in keeping a
large number of milk cows, and there is a great dairy called
the *Cascina*, which supplies the court at Florence with
butter; part of it is sold in Pisa, and is the best that can be
got. We have it sometimes, but it is not very remarkable.
There is another Cascina, *the Cascina Vecchia*, where a
number of camels are employed as beasts of burthen; I went
one day, and saw about twenty of them returning from their
day's work, laden with wood—they knelt down at the word of
command to be unloaded, and went quietly and soberly into
their stable, where they were fixed up like horses, to a rack
filled with hay. They are of a dun colour, with one hump on
their back, and altogether a very ugly animal. There is a
district on the sea-shore not far from the mouth of the Arno,
where these animals exist in considerable numbers in a wild
state, and where they have continued to breed, since the time
of the Crusades, when some were brought from the Holy
Land, by a Prior of Pisa. Those that are exhibited in the
shows of wild beasts throughout Europe, all come from this
place, where they are bought for six or seven pounds. The
Grand Duke has here also a great establishment for the
breed of horses.

We have been very fortunate in finding here, established
for the winter, Mrs. Drewe and her family, and the Miss
Allens. They are a very agreeable family indeed, and have
been very kind to us. I am quite domesticated there, and
spend all the time I can spare from my important duties at
home, in their house. Marianne and Georgina Drewe are

two charming girls, and they sing delightfully—exceedingly
well-mannered, gentle, and intelligent, quite the genuine
English manners. I have seen most things here in their
company, and you may conceive how much more agreeably
the sights are seen with such companions. They are come
abroad on account of the ill health of their sister Louisa, and
their brother. Louisa is a very· interesting gentle creature
about fifteen, whom it is impossible to look upon without
feeling a great interest in. They have come with the intention
of being three years abroad, and talk of spending next winter
in Rome. I shall be truly happy to have an opportunity of
making you acquainted with these excellent people, and when
we go to Devonshire, we have another family ready to give us
a hearty welcome. When I return to London, I shall, on'the
strength of their intimacy, renew my acquaintance with their
agreeable sister, Mrs. Gifford, with whom, you will recollect, I
was so much charmed last summer at George Nicholson's.·

On Christmas eve, after spending the first part with the
Achards, we went to the Cathedral about eleven o'clock, to
see the grand ceremony and bear the fine music of the service
on·that occasion. In both we were a great deal disappointed.
The high altar was lighted up with a great number of thick
wax tapers, and all the other altars in the Cathedral were
lighted up, but these were not sufficient to overcome the mid-
night darkness of so large a building; there was a prodigious
assemblage of priests in splendid dresses ; the Archbishop of
Pisa was seated on his throne with his mitre on his head and
chanted part of the service. The vocal music was very indifferent, but we had Corelli's service for the nativity, which
was very beautiful indeed. The effect of the ceremonies was
lost upon me in a great degree, by not understanding what
they were about, but there was very little impressive in it,
from the infinite changes and shifting of the scenes, some of
which were quite ludicrous. Nor did the effect seem to be

very remarkable on the Catholic congregation. There has
been a constant ringing of bells since 10 o'clock on Christmas
eve, how long it will continue I don't know, but it is a great
nuisance. The people are going about their occupations as
usual, and though on Christmas Day most of the shops were
shut, I saw butchers' stalls open in the market at mid-day,
and the country people selling their game, poultry, eggs, etc.,
as at other times.

I was entrusted by Lady Holland with a parcel for the mother
of an Italian servant (Raffaelli) who lives at Holland House.
His mother lives at Lucca, about twelve miles from hence, and
last Thursday I went there. As it would have been very dismal
to have gone alone, I asked the Miss Drewes to go with me, and
we made a very pleasant day of it. The drive is most beautiful
even at this time of the year, the road winding among the
hills of San Juliana, and for a considerable way among the
olive trees. Several villages and country seats of the rich
Pisans are scattered upon the hills, and there are two
castles situated on insulated hills, commanding two passes
of the valley. They are now in ruins and are more
picturesque perhaps on that account. The fields are all
cultivated with great care, and appeared in the best order.
We noticed several fields of lupins. The road passes for some
way along the Lerchio, a clear stream of blue water, very
different from the red muddy Arno. The Lucchese territory
is prodigiously populous, being six times as many persons to
the same space as in France, but it chiefly consists of land
that yields an abundant return, and is of very small extent,
being about forty miles long and fifteen broad. There are
great emigrations from Lucca to every part of Europe, and a
great proportion of itinerant glass blowers; the thermometer
makers, &c., are Lucchese. They are a very industrious
people, and exceedingly well dressed, and I saw many that
were very good looking; a gentleman who lived for some

months among them last year, told me that he found them remarkably intelligent, and very few of them who could not read and write. I must defer the farther account of my visit to Lucca for the next sheet.

Yours most affectionately,

L. H.

———

To His Sister.

Pisa, 31st December, 1816.

My DEAR NANCY,—Sunday was a very fine day, and Frank got a walk for an hour and a half, which did him a great deal of good, and the good effect of it was increased by his having the amusement of two ladies walking with him, Mrs. Drewe and Miss Allen, for although speaking is an exertion, yet if that is not carried too far, the inconvenience is more than counter-balanced, I think, by the amusement.

Yesterday was a raw, cold, disagreeable day, when he could not have gone out with safety; this morning it has been raining, but it seems as if it would clear up, and I hope we shall have sunshine enough to enable him to get out, for he depends more upon this exercise than anything else, for feeling tolerably well the rest of the day. . . . Miss Drewe, who had not seen him for a fortnight, told me she thought him looking better, when he walked out the other day. . . . . . In the last Florence newspaper there is a circumstance mentioned relative to Canova, the great sculptor, which does him infinite honour. The Pope some time ago, conferred upon him the title of Marquis of Ischia, and gave him at the same time an annual pension of three thousand crowns. This pension he has appropriated to the following purposes :—

1st. To the Antiquarian Society of Rome six hundred per annum.

I

2nd. A prize of a hundred and twenty crowns every three years to the most distinguished young artist of Rome, or the states of the Church, in the three branches of painting, sculpture, and architecture, and—

3rd. To those young men who gain the prizes, an allowance of twenty crowns per month for three years, when the prizes are again to be contested.

4th. To the Academy of San Lucca for the purchase of books on the Arts, a hundred crowns per annum.

5th. To the Academy of *Lincei* * a hundred crowns per annum.

6th. The remainder of the sum, one thousand crowns per annum, *per soccorire gli artisti domiciliati in Roma, che sono poveri vecchi ed inabili.*

This is an act of magnificence not to be met with in any country, or in any station, but very rarely. . . . . .

To-morrow is a *Giorno di Festa* for me, as my dear little Susan completes her first year. I hope in God she is quite well, and that there will be a merry meeting in Charlotte Square to-morrow. Give our kindest love to all. there, and I depute you to give a loving kiss to my dearest Anne and our three beauties, and if I am very envious of you, you will, I am sure, excuse me.

God bless you, my dearest Nancy,

Your very affectionate,

LEONARD HORNER.

Postscript by Francis Horner.

Many, many returns of the season to you, my dear Nancy, and to all at home. It is a long way to send such compliments, but they will arrive before old style is past. Leonard is remarkably well, and begins to find leisure to see a little of such sights as this place affords; he is now in the Campo Santo, studying the pictures, which, I believe, is the greatest

* See Note at the end of this Chapter.

and most valuable curiosity that Pisa affords. There are books published giving an account of it, and engravings indeed of the succession of frescoes, which form à sort of history of the art of painting in Italy, from its revival to its days of perfection. Those engravings may have reached Edinburgh, they were in London long ago. We shall have an opportunity soon of sending you a little book printed by a Professor of this University, which will give you some idea of the Campo Santo.

I have had a little walk, which always makes me feel better, cheering my spirits, and giving me a feeling too of improved strength.

My kind love to you and all,

Most affectionately yours,

FRA. HORNER.

Journal.

Pisa, *January* 18*th*, 1817.

I have found all classes here remarkably good-natured and obliging. There seem to be a vast number of lazy, idle people in · the town, but the peasantry round are very industrious. In my rides I have often talked with them, and generally found them intelligent and communicative, pleasant in their address, and very respectful, generally standing with their hat in their hand while you are speaking to them, and as you pass along the road, both men and women wish you a good day,—"Felice giorno, Signor," or "a Rivederla." In the town there are swarms of wretched objects, but the peasantry are all well clothed and healthy-looking. In our various rides we have gone upon every road that branches from Pisa within a radius of five miles; all this is in the flat alluvial country; the hills about five miles off must afford many charming walks.

The prettiest drive in this flat country is to the Duke's

farms, which stand in a park of vast extent; there is a thick
turf upon which you drive, and an oak forest of some miles
extent, through which a carriage can also pass. In the park
there is a very long avenue of the Pinaster; the trees are
young, but of a most beautiful green. There are, besides, a
great many Ilexes, very handsome trees, which at this season
look particularly well. It is, you know, the evergreen oak.
I observed, also, one large cork tree, which is also evergreen.
When we rode through the park a few days ago, it was
covered with white and lilac crocus. Daises have been in
full blow for nearly a month.

To His Daughter.

Pisa, 14*th January*, 1817.

MY DEAREST MARY,—I have not had a greater pleasure
since I left home than that which I received from your very
nice letter last Friday, and this morning your uncle was
made happy by receiving a letter from you. I am glad
you liked those letters of Lord Chesterfield's that your
mamma picked out for you; he writes very beautifully. I
think you must have found Forsyth's "Italy" not very
entertaining, as it is more a book to refer to when you are in
the places of which he is speaking, than to read like a book
of travels. . . . .

I have been looking about ever since I have been here for
some dolls dressed in the fashion of the country, but I have
not yet seen any. I should like to send you one in the dress
of a monk; there are a great many of these walking about
the streets of Pisa, and this morning, while I was walking out
before breakfast, I passed by a monastery, from which several
came out. They were Capuchins. Their dress is a dark
woollen gown with a large cape, or rather short cloak,
over it of the same stuff, and a hood which they draw
over their head when it is cold; the head is shaven quite

close, but some have a black silk or worsted caul that
fits quite close to their head; they have long bushy beards,
but their upper lip is shaven; round their waists there
is a rope, one end of which hangs down nearly to the
ground, and is knotted; with this, I suppose, they beat
themselves—you will recollect the the account of these
flagellations. in Townsend's " Travels in Spain." From their
waist also hang their beads, with a wooden cross at the end
of them.   They have bare legs and sandals upon their feet.
They are an idle and useless set of men,. and ought not
to be permitted to exist.   They were all scattered, and the
monasteries were abolished, when the French were in possession of this country, but since the return of the old
Government they have been re-established.

There are three or four of these monasteries in Pisa, and
also several convents. . . . .

Your uncle sends his kindest love, and he will write to you
very soon.

Farewell my dear, good little girl. I send you my
affectionate love, and my blessing to yourself and your
sisters.

Your very affectionate father,
LEONARD HORNER.

[On Thursday, the 6th of February, Francis Horner found
the difficulty of breathing increased, and his cough re-
appearing with some severity ; however, these bad symptoms
were somewhat abated on the morning of Friday; but
towards evening they returned, accompanied by drowsiness.
His brother slept in an adjoining room, and the dôor was
open between the two bedrooms.   In the night he heard him
moaning, so Leonard got up and asked him if he were
suffering.   Francis replied that he moaned from a difficulty
of breathing, but that he wished to be left to sleep.   Leonard
sent for Dr. Vacca, who came at seven in the morning.   It

was Saturday, the 8th of February. He found his patient
labouring greatly in his breathing, with strong palpitations
of the heart and irregular pulse; his forehead covered with a
cold perspiration, and his face and hands of a leaden colour.
He was, however, perfectly sensible, and spoke in a clear,
distinct manner, expressing neither apprehension nor anxiety
about himself. Various stimulating applications were tried,
but they afforded no relief. Although Leonard had entire
confidence in Dr. Vacca, still he felt it would be more
satisfactory to have further advice, so without consulting
Francis, he requested Dr. Vacca to arrange a consultation
with another physician. The two doctors came at four
o'clock in the afternoon, while Leonard was sitting by his
brother's bed. He went to meet them, and was absent
about ten minutes, and then returned to prepare Francis
for receiving the other physician, on drawing aside the
curtain he found Francis deadly pale, his eyes fixed, and on
touching him he found his hands cold. For a few minutes
Leonard fancied his brother had fainted, but when he found
that he was really dead, the shock was so great that he fell
down in a swoon. His kind friend, Mrs. Drewe, was soon
sent for, and gave all the comfort she could bestow on him.
The shock was the greater, as Leonard had no apprehensions
of this fatal end. His brother's cheerfulness and activity of
mind, and his being entirely free from anxiety about himself,
had deluded him into the belief that spring, and more genial
weather, would restore him to health. The physicians, too,
had not led him to anticipate such a sudden conclusion.
What added to his grief was the thought of the affliction of
his father and mother and sisters, when the fatal news
should arrive. He wrote to his friend, Dr. Gordon, in
Edinburgh, to break the sad tidings to his parents of the
death of their beloved eldest son, and the crushing of their
hopes and pride in the brilliant career that lay before him.

Mrs. Leonard Horner, who, with her children, was then living with them in Charlotte Square, described the agony of their grief; and on her entering the room, Mrs. Horner (the mother) came up to her saying, "It is *not* Leonard," so thoughtful was she in the midst of her grief. Time and a religious trust in the wisdom and goodness of God, brought resignation to these afflicted parents.

To return to Leonard, this grief was the bitterest he ever experienced, except the loss of his wife forty-five years afterwards, then he was an old man of seventy-seven, but at this time he was only thirty-two, and it gave a graver tone to his whole character. He wrote in his journal on revisiting Italy, on the 7th of October, 1861:—"Here I am again at Pisa, after an interval of within a few weeks of forty-five years. This place is associated in my mind with the greatest calamity of my life. I lost my father and mother when they were advanced in years; our dear boy when very young; but my brother Frank I lost when he was in the midst of a career of honour and usefulness."]

From Mrs. Drewe to Mrs. L. Horner.

Casa Caneggi, Pisa, *February 11th*, 1817.

DEAR MADAM,—Before I enter into the painful but interesting details I have to communicate, I must first assure you that Mr. Leonard Horner is in perfect health, and equal to the sad trial he has been called to; his sufferings have been acute, as must have been expected, but the first burst of sorrow over, he has so kindly yielded to every suggestion offered to soothe him, that I trust in God you will have nothing to apprehend in the shock he has sustained; he is now perfectly calm and resigned, and he will tell you so himself, by the same post by which I send this letter. He is a good deal occupied this morning, and may not have time to give the ample details he would wish. It will, I hope, be a comfort to you, and Mr. and Mrs. Horner, to know that your dear husband is with friends entirely interested in his sorrows, and who will watch over his health and spirits, and who have had the happiness of yielding him perhaps, as much comfort as he could know (removed at such a distance from his own natural friends) in the affectionate sympathy with which we share his sorrow. We loved him first for his brother's sake, we now love him for his own, and I am sure from his laudable exertions and fortitude, he will reward us by permitting us to restore him to you with unabated health, and with spirits subdued, but not broken; which the sight of you and his dear children will again recal to their wonted tone.

With you, dear Madam, and his afflicted family, the nation will long mourn, and will share in this great subject of your domestic grief. We have lost our chief support, our fairest promise—the comfort that man cannot give can alone whisper comfort, when time shall have mellowed grief into calm and tender recollection. I trust it will be a satisfaction to Mr. and Mrs. Horner, and his family, to know that no stranger was permitted to do the last offices required. The calm serenity of his fine countenance was unaltered, and bespoke the uncommon ease with which his great spirit departed. I sat by him till all was closed. To-morrow I accompany Mr. Leonard Horner, together with the rest of my family, to pay the last respect at the sad ceremony. I have no doubt Mr. L. Horner will not disappoint my reliance on his firmness. I shall not lose sight of him till he leaves Pisa finally, and will place him in the carriage that conveys him to you.

With the united best wishes of my sisters and daughters, who cannot feel as strangers to you, excuse the liberty I take in subscribing myself, my dear Madam, your sincere friend,

And obedient humble servant,

C. DREWE.

From Lord Lansdowne.

Naples, *20th February*, 1817.

MY DEAR SIR,—I have only received this day your kind letter of the 14th, which found me, as I have been since the moment I heard of it, and must long continue, in the deepest affliction at the melancholy event to which it relates. Your brother's loss is indeed irreparable, not merely to his family and friends, to whom the purity of his mind and affections of his heart had always endeared him, but to the public, whose interests his admirable talents and highly-cultivated understanding were so well adapted to promote; for myself I can truly say, that as the only pain that could arise out of a connection with him was that of losing him, so that pain has proved one of the greatest I have ever experienced, and one from which I in vain seek relief from the distractions which the beautiful scenery in the midst of which we are now residing affords. His memory, and everything connected with it, I shall ever cherish, with the warmest feelings of affection and regard.

All the consolation which the sympathy of many real and valuable friends can supply, you will not fail to derive; more you can find in time only, which alleviates, though I am sure in this instance it can never efface it. I grieve to think of that which your father and mother must experience.

Believe me, both on your own account, and on account of him whom we have lost,

Yours ever,
Very sincerely,
LANSDOWNE.

From Lady Holland.

Holland House, *7th March*, 1817.

I am persuaded you will ascribe my silence to the true cause. I have several times begun writing, but have not felt equal to the task, and I have also twice attempted a letter to

Edinburgh. My only anxiety now upon the subject is in case
you should arrive at.Paris without finding a line from any
of us, which will be the effect of an unpardonable and
selfish indulgence of my own feelings, instead of sufficiently
considering yours. I have been ill and so totally overcome,
that we have this day returned to Holland House, having
given up London entirely. I hope the change may restore
my spirits and enable me to sleep better than I have hitherto
done. I trust you will give a day, or as many as you can
spare from your business in town.

Most truly and affectionately,

Your sincere friend,

E. V. HOLLAND.

[Francis Horner was buried on Wednesday, the 12th
February, in the Protestant Cemetery at Leghorn. His
father erected a monument over his grave. It was designed
by Sir Henry Englefield, and on one of the sides of it, there
is a medallion of the head of Francis Horner in profile,
executed by Sir Francis Chantrey.

Mr. L. Horner had many arrangements to make at Pisa,
before he could begin his sad journey home. He did not
reach London till after the first week in March. Here, he
stayed a few days to see Lord and Lady Holland, and some of
his own, and his brother's friends, who not only sympathised
in his affliction, but shared in his grief. Before he arrived in
England, on Monday the 3rd March, there was a motion in
Parliament for a new writ for the Borough of St. Mawes, now
vacant by the death of Francis Horner. On this occasion,
a most touching and honourable tribute was given to his
virtues, and the heavy loss the country had sustained by his
death, by members of all parties—a circumstance, I believe,
almost, if not quite unprecedented before and since, and most
gratifying to his family. When Leonard arrived in Edin-

burgh, his wife and parents were shocked at his altered appearance, from the sorrow and fatigue he had gone through since the 8th of February.]

From Mr. Allen.

Holland House, *March* 19*th*, 1817.

DEAR LEONARD,—In case you have not heard directly from Edinburgh, it will be some consolation to you to know that your father bears his misfortune with more firmness than was expected, and that your mother and sisters, though in the deepest affliction, submit to it with resignation. You must have suffered much, and had a painful duty to perform, but it will be a satisfaction to you for the rest of your days, that the last months of so valuable a life were passed in your society, and that to your care and attention he was indebted for all the comforts that relieved and softened the last period of his existence.

I have seldom seen the fate of any man so universally lamented as your brother's. All parties concur in extolling his merits, and in lamenting his loss. All agree that he was a person to have risen if he chose to the first situation, legal or political, in the kingdom. The tribute to his memory in the House of Commons, was most gratifying to all his friends, and I think must be equally so to all his family. Nothing could be better said than the enconium of Lord Morpeth ; Manners Sutton spoke under the impulse of strong feeling, and without premeditation, and Romilly's eulogy on the independence of his character, came with the best possible effect from him. The few words of Lord Lascelles, showed how unanimously the House felt on this occasion. A corrected report of what was said, is to be printed, and if ready before the post goes on Tuesday, it will be sent to you.

Lady Holland, as you may well imagine, was deeply affected by this sad and most unexpected misfortune. They have left town, and are now at Holland House, where they will expect to see you as soon as possible after your arrival.

Yours faithfully,

JOHN ALLEN,

To Dr. Marcet.

Edinburgh, *October* 14*th*, 1817.

MY DEAR MARCET,—I have read with the greatest sorrow in the *Morning Chronicle* to-day the death of Mr. Haldimand. When I last saw you, you were in the hope that his health was so far restored as to remove that state of constant apprehension in which you had been so long kept. It is very lamentable to see the *key stone* of such a family carried away, for it would be contrary to all experience, I believe, were the materials to remain long united in the way they have been. Mrs. Marcet has been long prepared for this separation, but to be deprived of the object of her care must afflict her greatly. Pray let us know how she is, for we enter with great earnestness into all the sad feelings of the family at this moment. Tell us also how Frederick is. We got home on the 5th, and had the happiness of finding all our children perfectly well, and my father's family in tolerable spirits.

Give my very kind regards to Wm. Haldimand.

Believe me, my dear Marcet,

Your with great regard,

LEONARD HORNER.

Edinburgh, *December* 13*th*, 1817.

MY DEAR MARCET,—As you are very likely to seek relief from your sorrow * in active exertion, I do not hesitate to ask you to take a little trouble for me. Some circumstances

* The death of his boy Frederick.

have come to light of gross mismanagement in the domestic economy of the Infirmary here, which will, in all probability, occasion a committee of contributors to be appointed to inquire into the abuses. I am one of those who are taking a little trouble to investigate the matter, and in order to discover in what points it is defective, we are endeavouring to get information with regard to the management of other hospitals throughout the country. I understand that Guy's is one of the cleanest and best managed in England, and if there are any printed regulations respecting the management of it, I shall be obliged to you to send me a copy. I hear also that the London Hospital is well conducted, perhaps Dr. Yelloly can supply me with a copy of the regulations of it.

The chief causes of complaint in the Infirmary are, the great filthiness of the patients' beds, the scantiness and bad quality of their food, and the neglect of the nurses, who are often drunk. The medical department is, from all I hear, on a very good footing, as far as the patients are concerned. It is a most important thing that the Infirmary here, which is so material a part of the medical school, should be on the best plan, and should keep pace with the latest improvements in every hospital throughout the world. On the contrary, very little alteration has taken place in the domestic economy of it for forty years, and the medical officers have remonstrated in vain. The constitution of it is bad, and leaves the whole management in the hands of a treasurer, who will not allow himself to be interfered with. If we succeed in effecting any improvement, it will be entirely owing to the steady exertions of some Quakers,—tell William Allen this.

My friend, Dr. Gordon, is greatly pleased with your work, which he says is " admirably done."

Give Mrs. Horner's and my kindest regards to Mrs. Marcet; pray write to me soon and tell us how you all are—a few lines will do. I was delighted with the kind intimation Mr. Wm. Haldimand gave me about Prévost. Some people are born with golden spoons in their mouths; it is found out at last, although for some time they fancy it made of horn. It is hardly fair, however, to apply this saying to the present case; few are so capable of converting horn into gold as Prévost is.

I hope Frank and your two little girls are well. Mary sends her love to her friends. Remember me very kindly to Roget.

Believe me, always, my dear Marcet,

Yours with sincere regard,

LEONARD HORNER.

NOTE.—The Lincei was founded by Frederick Cesi. He was born at Rome in 1585, and was the son of the Duke of Acqua Sparta, a family known only for their pride and ignorance. But nature had created in Cesi a philosophic mind, he gave his entire regard to Science, and projected himself at the age of eighteen, an Academy, that is a private association of friends, for intellectual pursuits, which with reference to their desire of piercing with acute discernment, into the depths of truth, he denominated the Lynxes. Their device was that animal, with its eyes turned towards heaven and tearing a Cerberus with its claws; thus intimating that they were prepared for war against error and falsehood. The Church, always suspicious and inclined to make common cause with all established tenets, gave them some trouble, though neither politics nor theology entered into their scheme. This embraced as in other academies, poetry and elegant literature, but physical science was their peculiar object. Porta, Galileo, Colonna, and many other distinguished men both of Italy, and the Transalpine countries, were enrolled among the Lynxes, and Cesi is said to have framed rather a visionary plan of a general combination of philosophers in the manner of the Pythagoreans, which should extend itself to every part of Europe. . . Cesi died in 1830 and his Academy of Lynxes did not long survive the loss of their Chief.—*From Hallam's Literature of Europe.*

CHAPTER VI.

1818-1819,

Leonard Horner to his twin-sisters on their birthday.

London, *February 12th,* 1818.

My Dear Fanny and Nancy,—Although I have not time to say all I wish to you I cannot allow this day* to pass without sending you my affectionate love and congratulations upon its return. It will not do to think on my occupation on this day last year,† I must look forward and hope that happier days are to come.

I spent a very pleasant day at Lansdowne House yesterday; there was a quiet sociable party of seven. Lady Lansdowne is greatly improved to what she was when she returned from abroad, indeed she has recovered her former beauty. She is a very agreeable person. Lord Lansdowne has brought from Rome a very fine thing by Canova. It is a Venus nearly the same as his statue in the Grand Duke's Palace at Florence, which is so celebrated. It is exquisitely beautiful and is placed very advantageously in the dining-room (which is a very fine room) before the window, and when the scarlet window curtain is let down, and the light from the centre of the room shines upon it, the effect is very striking.

Lady Lansdowne was at Chantrey's yesterday for the first time, and she was very much pleased; indeed she spoke in stronger terms of the bust‡ than anyone I have seen.

* Their birthday.   † Francis Horner buried at Leghorn.

‡Bust of Francis Horner, now in the possession of his great-nephew, Leonard Lyell, of Kinnordy, N.B.

Many, many thanks for your kindness to my dear children. I can well imagine Mary looking in her new frock such as to make me very happy.

Give my kindest love to all at home.

I am ever most affectionately yours,

LEONARD HORNER.

Edinburgh, *April 4th,* 1818.

MY DEAR MARCET,—The Infirmary Committee have terminated their labours, and on bringing up their report, which was in a tone very far from being agreeable to the managers (though much under what the evidence would have borne out), they, in the true spirit of their political creeds, brought such a host of voters from every part of the country, that by a great majority it was carried that *black is, and always was, white.* However, the evidence has been published, so that the good citizens of Edinburgh will be able to see how their hospital has been managed—but what is more material, such a radical reform has taken place in the Infirmary since the appointment of the Committee, that I am told by some of the students, it is no longer like the same place.

I believe you are not acquainted with my excellent friend, Murray; the account I can give of him is that he was the most intimate friend of my poor brother, from the age of four to the day of his death. He is now in London, and would, I know, be very glad to meet you and Mrs. Marcet. He will see a great deal of Whishaw, in all probability.

I have seen Captain Hall several times, and thank you very much for introducing me to his acquaintance. He is, as you may well suppose, greatly in request. I dined at Sir James Hall's on Friday last; I was grieved to see so sad a change in the looks of Lady Helen. I had not seen her since she has been visited with so many dreadful misfortunes, and her altered looks too plainly show how deeply they have been felt by her; her spirits were, however, very good, and she seems to be very, very happy with her son. Sir James is better, but he is evidently labouring under a disease that has an effect upon his mind. There is an indistinctness in his memory, and a confusion in his ideas and articulation, which is very distressing. It is quite beautiful to see the amiable attentions with which he is constantly watched by his son Basil, he is a very edifying example of filial duty and affection. He talks of leaving Edinburgh in ten days for London, and of making a tour into Switzerland and Italy in the course of the summer.

Playfair is to read at our Royal Society next Monday a Biographical Memoir of Clerk, the author of "Naval Tactics," I hear it is a masterpiece. Leslie brought forward a new instrument at the last meeting, called an Æthrioscope; there is an account of it already given in the part of the supplement to the Encyclopædia Britannica just published, in which you will find a great deal of Leslie's ingenuity, but Brewster says that he has been anticipated in all the essential points of his theory by Wells in his essay on Dew. Jeffrey is busy with a new number of the *Edinburgh Review*, and I believe it is Gordon's wish, if he can possibly get it ready in time, to bring you forward in this number.

Mrs. Horner and our little flock are quite well, she has been made very comfortable by the good accounts she has received of her sister, who seems as if she would cheat the doctors.

Give our very kindest regards to Mrs. Marcet, and believe me, my dear Marcet,

Yours with great regard,

LEONARD HORNER.

K

To Dr. Marcet.

Edinburgh, *May 21st*, 1818.

My DEAR MARCET,—I have put off writing to you so long, that I fear my letter will hardly reach London before you set out on your journey, for I think you talked of leaving London about the beginning of June. And yet, I hope that some accident may delay you a *little* longer, for then I might chance to see you, as I expect to be in London in a fortnight or three weeks. You will have a very delightful journey, I cannot conceive one that would promise more happiness.

Murray is returned here, and told me that he regretted extremely that he could not accept your kind invitation. He had the pleasure of meeting Mrs. Marcet at Lady Davy's. He says he will make a point of calling upon you among the first places he goes to, when he is next in London. I suppose you have seen Captain Hall by this time, and probably arranged some plan of travelling together. He seemed very anxious to bring this about. Brewster you have probably seen by this time. . . . . His kaleidoscope seems to have amused the Londoners amazingly. It is a very pretty toy. He has, like a true philosopher, managed this matter badly as an article of trade, for William Allen, who was here lately, told me that if he had managed matters better, he might have made three or four thousand pounds by it.

Mrs. Horner is very well, but she does not go with me to London, because my stay must be short. Remember us in the kindest manner to Mrs. Marcet, and all our kind friends that we used to meet with such pleasure round your fireside.

I am ever, my dear Marcet,

Most faithfully yours,

LEONARD HORNER.

P.S.—You will be sorry to hear that Dugald Stewart has had an alarming attack of fever, he is now past all danger, and is indeed recovering fast.

To His Mother.

London, *June 24th*, 1818.

MY DEAREST MOTHER,—I had the pleasure on my return from Little Bounds yesterday, to receive your letter. You are very kind, my dear mother, in proposing to give us a carpet for our new dining-room. After much hesitation as to the propriety of going to the expense of a Turkey carpet, I at last persuaded myself that it was the most economical, and went last week to the Levant Company's warehouse, where I got the one now in my dining-room, and bought one which was sent off last week. You speak of our prospects in our new house. I cannot look forward to any enjoyment without a strong feeling of the uncertainty of all such hopes. I have many, very many blessings to be thankful for, and should be very ungrateful to the Giver of all good, were I to murmur at my having shared in the sorrows of life, but so many sad changes have occurred of late in our family * that it is not surprising if my confidence in the future is not very great. It is a very difficult thing to acquire that temper of mind, to enjoy with due cheerfulness, the good gifts of fortune without forgetting how uncertain their tenure is. I have attended to your commission about your kind present to Mary and Frances; as it is desirable that such a present should be useful and often in their sight, I have bought a silver chamber candlestick for each, but Mary's is the handsomest of the two. What a blessing it is to my dear Anne and me to see the affectionate interest you all take in our children. I trust they will ever shew a due sense of it when they are more able to appreciate its value.

The town is in a great bustle about the elections, it is considered a prodigious triumph that Curtis has been thrown out, and it is a very remarkable proof of the change in

---

* The recent deaths of his wife's father and sister.

popular opinion, for he has hitherto always come in at the
head of the poll. Maxwell it is now thought may be turned
out, but there is some uneasiness in Romilly's friends, for the
Burdett party are behaving excessively ill, are violent against
Romilly, and are giving plumpers to Burdett. It is a great
thing too, that Barclay has been turned out from the
borough.*

Give my kindest love to my dear father and sisters, to
Anne, and all my sweet brood. Tell Anne that I shall write
to her to-morrow.

Farewell, my very dear mother, and believe me always,

Most affectionately yours,

L. H.

From Dr. Marcet.

Geneva, 8*th July*, 1818.

MY DEAR HORNER,—With the exception of Prévost, you
have been the first of our friends to write to me, and I feel
exceedingly obliged to you for your kind remembrance.
Besides, you really were the first to give me all the news that
interested me the most. I was thunderstruck at the news of
poor Dr. Gordon's death, a most excellent man, from all
accounts, and of a rare species, particularly in Edinburgh,
where talent and independence are not often united. I have
felt for you sincerely.

On arriving here, three weeks ago, we alighted on a very
pretty country house on the western bank of the lake,
and within a mile of the town. The weather was heavenly,
and Mont Blanc, presiding majestically over the Alps, just in
front of our terrace; Frank was there to receive us, with my
sister and her excellent husband, and three plain, but honest
and zealous Swiss servants, had been previously placed in

* Sir Samuel Romilly carried his election by a great majority.

the house by my sister's kindness, so that in less than two
minutes we were completely at home. I found my mother
grown old, but not infirm, and perfectly delighted to see the
only granddaughters she ever had. From the first evening our
friends began to pour in from all quarters, and the tide has
yet hardly subsided. We go out every day, either to dinner
or to ride, or upon the lake, and we are all in perfect health.
This is happiness indeed, but in this world of vicissitudes,
how long will it last? I lead the idlest possible life, and yet
I do not experience any desire of change, at least at present.
In October, however, all this must be over, and Russell
Square and Guy's Hospital will again be the order of the day.
But I have made up my mind to quit private practice, which
will improve my professional enjoyments very much, for I
have always found that all practice, except that of the
hospitals, operated as a sad interruption to my state of
mind and to my favourite pursuits. I have found Frank
improved in size and steadiness, and he is perfectly happy.
He is completely under the control of Geneva customs and
Geneva opinion, the only safe-guard for young Englishmen
at Geneva; for amusements are so incessant here, especially
for young English people, and so easy of access, that
without discipline a young Englishman is very soon lost—
not however, to virtue or morality, but to study and labour.
I have found our Government grown a little bolder, and
certainly less in the *ultra line* than they were three years
ago. The perfect equality of ranks here, and comparative
simplicity of manners, are quite delightful to me. In this
respect nothing can exceed the independence of spirit of the
Genevese. This place is full of English lords and German
princes, but no attention is paid to them by society, except in
proportion as they are agreeable and well-informed. The
ambition of Geneva is at this moment more especially
directed towards the extension and improvement of its

establishments of education, and I hope it will be with effect.
Young Necker is about to publish his travels in Scotland, and
is attending more to geology than he had done for some years
past. He has, however, just been appointed Deputy to the
Diet, which will occupy him two or three months. Adieu,
pray let me hear from you again. I say nothing of the
catastrophe in Valais, as I know the public prints have been
very full on that subject. I strongly advise Necker to read a
scientific account of it to the Geological Society.

Mrs. Marcet joins in kindest remembrances to you, and
Mrs. Horner.

Ever most sincerely yours,

A. M.

From Lord Lansdowne.

Bowood, August 28th, 1818.

DEAR HORNER,—It gave me great pleasure to see you have
had an addition to your family *, and I should be very glad,
when you have a moment's leisure, to hear how Mrs. Horner
does, and indeed how you yourself do, for I think you were
by no means well, when you were last in London.

Lord Auckland writes me word that since I left town,
Chantrey had finished his design for the monument† we are
all so much interested about, which I could not get him to do
before, he approves of it very much, but we shall have a
meeting in the winter before it is finally settled, and this will
make no delay, as I am afraid Chantrey has too many orders
to expect him to proceed with any rapidity; he has lately
received one from New York for a statue of Washington, while
Canova has received a similar one from Philadelphia, which
flatters him very much.

---

* The birth of a fifth daughter, afterwards married to Chevalier Pertz.
† The Statue of Francis Horner in Westminster Abbey.

We shall spend the remainder of the year here, where, I am afraid, we shall have little chance of seeing you and Mrs. Horner, but trust sooner or later you will find, or make an opportunity.

Believe me, ever yours truly,

LANSDOWNE.

To Dr. Marcet.

Edinburgh, *October* 16th, 1818.

MY DEAR MARCET,—It gave me very great pleasure to receive your kind letter from Geneva, and to learn that you and Mrs Marcet were enjoying yourselves so much ; I should have been delighted to have been breathing the same atmosphere with you, for to have seen you, as your letter enables me to imagine, I must have participated in your gaiety.

There was one subject I wished to write to you about, ... I mean the review of your book. There was nothing found in poor Gordon's papers relating to it ; indeed, when I saw him last, which was ten days or a fortnight before his death, he had written nothing, but was collecting materials. I spoke to John Thomson, and he is very anxious that it should be done, but it is impossible for him, with his other avocations, to undertake it. He called upon me a short time ago to ask me whether I would write the review of the chemical part, and he would keep me right as to the medical part ; I did not say that I would not, but upon further consideration, I feel myself quite unequal to do your book justice. . . . . . . I have had the pleasure of seeing Warburton here this season, and I assure you it was a great treat to me to see one of those persons in whose company I had spent so many pleasant hours, and the loss of whose society was the source of my regret in leaving London. He came to visit his sister, Lady Elphinstone, an agreeable person, with whom he has made us acquainted. He has been making a tour in the Highlands, with which he is much pleased, and he is now on the eve of his return to London. He gives us some hope that he will visit Edinburgh again, and I trust he will come when some of the people most worth seeing are in town. I have also had the pleasure of renewing the acquaintance I formed with Captain Kater, one day at dinner at Dr. Wollaston's.

Since I wrote to you, poor Mrs. Horner has had some very severe afflictions to bear. About two months ago her mother died, and yesterday brought us the melancholy tidings that poor Mrs. Winthrop's sufferings were at an end. Thus in fourteen months she has lost her father, mother, and only sister. Her mother's death occurred at a time which gave me great alarm, for she was within a few days of her confinement. She got over that, however, very well, and she gave me a fifth daughter, who thrives as well as we could wish. She bears up against these misfortunes with great fortitude, and her health has not suffered. Poor Winthrop will feel his loss most deeply, and the situation of his two girls is very sad, as they have no near relative who can look after them.

Mr. Dugald Stewart is going to spend the winter in the South of Devonshire—he had a severe illness early in the summer, from which he has not entirely recovered, and it is thought prudent to take him away from his more severe studies.

Give Mrs. Horner's very kindest regards to Mrs. Marcet. We trust all your children are well. Mary sends her love to Louisa and Sophia.

Believe me, always, my dear Marcet,

Most faithfully yours,

LEONARD HORNER.

From Dr. Marcet.

London, *November 2nd*, 1818.

MY DEAR HORNER,—Your. very kind letter reached
London nearly at the same moment as myself. Everything
seemed to conspire to render our return melancholy. We
heard at the same moment of two deaths in your family, of
poor Lady Romilly's hopeless state, and of her brother,
Colonel Walsham, being nearly dying of a complaint very
like her own, in Sir Samuel's house. All this, contrasted
with the long protracted state of enjoyment in which Mrs.
Marcet lived, previous to her return, threw her into a low-
spirited state, which has continued ever since. Poor Samuel
arrived last night in his house in Russell Square, and desired
to see me two or three hours afterwards. I found him in a
distressing state of restlessness, both of body and mind, as if
worked up by a long and cruel state of suspense, and by too
great efforts of courage, into a sort of nervous *érethism* which
is most harassing. His general health, however, appears
good, though he does not think it so himself, and I hope and
trust that a gradual return to his habits and pursuits, will
gradually restore him to his usual energies. Dumont and
Roget were with him, and have not quitted him for some
weeks past.

Monday evening, half-past six. With feelings of the
greatest horror I have to communicate to you that poor
Romilly is no more. . . . . Sophy,
and Romilly's sisters are with us. William is at his uncle's.
They are all about us—and it affords us some consolation to
share in their sufferings. Roget is in a state truly worthy of
compassion.

Pray do not spread this dreadful news (at least publicly)
till you hear of it by the newspapers.

To Dr. Marcet.

Edinburgh, *November 5th,* 1818.

MY DEAR MARCET,—It was with feelings of the deepest affliction, that I read the dreadful news you communicated to me in your letter which I received this morning.

I had heard of the death of poor Lady Romilly, and dreaded, as many others did, the effect it might have upon Sir Samuel's health; but when I considered the man he was, I flattered myself that he would seek relief from his domestic affliction in the activity of public life, and that the country would not be long deprived of his invaluable services.

There is none now left to whom we can look up with veneration for splendid talents, high-minded exalted views, and unsullied virtue; and the country is doomed to sink, for bigotry, intolerance, and corruption will triumph without control. His loss to the public is what forces itself most strongly on our minds, but to turn to his wretched family, feelings of a different cast arise. What a sudden annihilation of the brightest state of existence! for his family had attained everything that could render life valuable, and most to be desired. It would not be possible to invent a more heart-rending tale. In the beginning of your letter you say Mrs. Marcet was far from being well, from the many melancholy circumstances which damped the joy of your return home. I fear that so sudden and violent a shock to her feelings, may have done her serious injury, and I beg that you will have the goodness, on receipt of this, to write me even half a dozen lines, to say how you all are. The poor Rogets, too! what a dreadful calamity to them; pray say to Roget how sincerely I sympathise with him in his affliction, and that I would have told him so myself, but for the impossibility of writing as I could wish on such a subject. I have used the first person in telling you the effect of this sad calamity, but there is but one feeling of sorrow in

our family, and indeed among those to whom, as personally
known to poor Sir Samuel, I felt it a duty to communicate
the contents of your letter.   What a blow it would be to poor
Whishaw, tell me how he has supported it, and how he is.

I can add no more at present, for I have written these
hurried lines, to beg of you to tell me how poor Romilly's
children are, and all those by whom his loss is most severely
felt, that you know about.

I am ever, my dear Marcet,

Most affectionately yours,

LEONARD HORNER.

To the Same.

58, Great King Street, Edinburgh, *November* 18th, 1818.

MY DEAR MARCET,—I wrote to you the day after I got your
letter with the melancholy intelligence of Romilly's death.   I
feel most anxious to hear from you, to know how you are,
for I am sure there are many circumstances that have
attended this calamitous event which must have disturbed
you greatly.   I pray, therefore, that you will send me a few
lines at your earliest leisure. . . . . We
have removed to a house in town, and are now in the
uncomfortable state of confusion which such an event must
always create.

Mrs. Horner unites with me in kindest regards to Mrs.
Marcet and yourself.

Believe me always,

Most faithfully yours,

LEONARD HORNER.

From Dr. Marcet.

London, *November* 26th.

I am much obliged to you for your kind enquiries respecting
the effect of the late most disastrous events in our neigh-

bourhood, on Mrs. Marcet's health and spirits. She at first
bore the shock much better than I should have expected,
which was evidently to be ascribed to the violent exertion
she made to support the courage of Miss Romilly. William
Romilly, and Mrs. Davies, whom I brought to our house
immediately after the catastrophe, are in a state which I
should in vain attempt to describe. They remained a week
with us, and that week was passed in a succession of scenes,
which you can better imagine than I could possibly relate.
These poor children, however, behaved with becoming
resignation and fortitude, Sophy Romilly is a very angelic
creature, and there is a character of sweetness in her grief,
which renders her peculiarly interesting. They are now all
united at Mrs. Davies's in Radnorshire, and derive great
consolation from mutual sympathy and affection. ˙ Sir
Samuel leaves upwards of one hundred thousand pounds,
which he divides among his seven children, leaving a double
share to his eldest son. He appoints Whishaw, and Lord
Lansdowne executors, and the former, guardian of his
children. He recommends to Whishaw's care and revision
his work on criminal jurisprudence, which was nearly
ready for the press, and some memoirs of his own
life, to which he seems to attach but little importance. The
children, and particularly Sophy, are quite delighted at
Whishaw's appointment, and Whishaw on his side conducts
himself as usual, with uncommon tenderness and judgment.
Poor Roget has suffered cruelly. For some days the image
of his uncle trying in vain to speak, was incessantly pursuing
him. He loses almost everything in losing his uncle,
but yet he bears this with becoming fortitude, and has
already begun to return to his usual occupations.

There is a young lawyer from Geneva attending some
of the classes in Edinburgh, Mr. Pictet, with whose father I
am very intimate. The said father is *Judge Supreme* at

Geneva, and a man of great merit. If you should meet this young man and show him some kindness, I should feel much obliged to you.

Believe me, ever,

Most sincerely yours,

ALEXANDER MARCET.

From Dr. Marcet.

London, *7th June*, 1819.

MY DEAR HORNER,—Before the *packing up* fever seizes me, I must acknowledge the receipt of your very kind letter, and tell you (which I can do with the most perfect sincerity) how keenly I regret missing your visit in July. But I hope you will write to me from London, in some degree to make up for this disappointment, and I promise to be as punctual a correspondent as I am capable to be, though I cannot aim at anything like the epistolary merits which I always find in your letters to possess. If you did not know that I am a *straightforward sort of fellow*, I should be afraid of being accused of making compliments; but I know you will allow me to be as free from that fault as any man, and the real fact is, that I always find in your letters what I value most—expressions of friendship which I know to be sincere; miscellaneous information which I know to be true; and a certain independence of sentiments which agrees perfectly with my own; and, besides, we have a curious resemblance with each other in our manner of receiving men and things, which is very grateful to me, and would allow quite a *rest* to our correspondence. I often thought of you and your ever lamented brother, during the late great victory of truth and principle in the bullion question. I never witnessed any public event in this country which pleased me so much. It seemed like the triumph of abstract truth and justice over the immediate interests and prejudices of almost all those concerned in this question. I am not quite a Tory yet, but I really think some of the Cabinet Ministers have behaved with the utmost candour and disinterestedness upon this vital occasion. What a pleasant thing it is to see the small minority in the Bank (two or three only) supported by Government, while the great majority of humbug directors, are at last thrown into the background! and what a gratifying circumstance it is to me that one of those few right-headed ones, should be a person to whom I stand so nearly related,* and that he should have had so great a share by his firm and luminous evidence to bring about that great event. How this posthumous triumph of your brother's labours must have delighted you and your father! If any feeling can be called delightful which associates itself with the loss of such a man.

Adieu, my dear Horner, we intend to be off on the 18th, so if you feel inclined to favour me with another letter, there is time. Mrs. Marcet begs to be most kindly remembered to you and Mrs. Horner, and I join in embracing all your family.

Ever yours,
A. M.

To Dr. Marcet.

Edinburgh, May 22nd, 1819.

My Dear Marcet,—I have been speaking to McCulloch about the "Conversations on Political Economy," and telling him how much good he would do, by making this excellent little work more extensively known. When I asked him if he knew it, he said "Know it? Why there it is on my table before me—I have read it three times regularly through, and am

* His brother-in-law, Mr. William Haldimand.

constantly referring to it.   It contains all the soundest and most enlightened views, most clearly laid down." He said he should certainly consider whether he could make a review of it to satisfy himself. . . . . You are now, I presume, in the midst of active preparations for your great journey.   I shall regret your absence most deeply, as cutting off one of the greatest pleasures I have during my now, alas! short visits to London.   But I hope you will write to me now and then.   Mrs. Horner sends her kind regards to yourself and Mrs. Marcet, and joins with me in wishing every possible prosperity to your journey. Remember me very kindly to Mr. W. Haldimand and to Prévost.

I have nothing new to tell you from this place.   No man is doing anything in science except Brewster, and I have not heard of anything very interesting that he has been about lately.   His new journal promises well.

I am always, dear Marcet,

Yours most faithfully,

LEONARD HORNER.

P.S.—Remember me most kindly to Mr. and Mrs. Mallet.

To Dr. Marcet.

Edinburgh, 13th June, 1819.

MY DEAR MARCET,—I will not allow you to leave England without my wishing you every possible prosperity in the very interesting journey you are about to undertake.   You go under the most favourable circumstances, and with every advantage for the full enjoyment of the many interesting places you will visit.   I most sincerely trust that nothing will occur to disturb for a single day the schemes of pleasure you have formed for this long absence from England.   All I ask is, that you may be more confirmed in your attachment to England, that you may have no doubt that the society of London, such as you have it, is better than any other in the world, and that you may feel some degree of impatience to see it again before your two years have elapsed.

I noticed your name in the list of presentations at the levee, and I concluded at once what your motive must be for going through such a ceremony. I daresay you took care to avoid the mistake of another philosopher of our acquaintance when he was putting on his sword to go to Court. I rather wish you had been presented by another.

I shall say nothing more on the subject of the reviews, than that I shall feel more anxious to have my wishes accomplished, as you are to be absent from England, for it would be so very agreeable to me to enclose some proof sheets, and I am sure both you and Mrs. Marcet will be pleased with whatever is sent, for there is a very strong laudatory tendency in both cases.

It has been, indeed, a very gratifying thing to our family to see the great triumph that has been so generously decreed to my lamented brother ; I say generously, for Peel's avowal of the superior sagacity of one who had been uniformly opposed to him, was made with great manliness, and without qualification.

You will recollect that I was in London when Mr. Haldimand was giving his evidence, and I then heard the great effect which his evidence had produced.

It was the general impression then, that no good would come of the Committee, but I recollect he told me that his conviction was that something would be done. He is now placed in the highest rank of the merchants of this country, and his example should teach other young merchants, that to attain the station he holds, something more is necessary than the ordinary vulgar routine of the desk. It is very comfortable to see the stupid conceit of your pompous bankers, who glory in their contempt of theory, brought down to its just level.

You have no doubt seen Brewster's new journal. He is very active, and as his foreign correspondence is very extensive, it will probably go on well for some time. He has, however, fallen into the error of all the journals hitherto published, by giving so scanty a place to his account of what is going on abroad. If while you are abroad you pick up any little bit of scientific news, which you think would be interesting, and would send it to me to give to Brewster, I shall be obliged to you. A letter from you now and then will be a great treat to me, and I shall not fail to write when I can tell you anything which may be interesting to you. We have many friends at Geneva, for whom we have a great esteem; pray give our best regards to the Achard family, and the Schmidtmeyers. You may tell the friends of Dr. Colladon and Mr. Coindet that they are very well, for they dined with me yesterday. Mrs. Horner sends her kind love to Mrs. Marcet with every good wish for the success of the grand tour.

With kind regards to Mr. Haldimand and Prévost.

I am ever very faithfully yours
Leonard Horner.

From Mrs. L. Horner to her Brother.

26th August, 1819.

Leonard arrived safely last Monday se' night quite well. Many thanks for your kindness to him during his visit to London.

On Tuesday Dr. Parr arrived, and from that time to this our thoughts, our time, and all our sense, have been occupied by the old doctor.

He brought two young men with him, and we being his

only old friends, have had everything to do. It is a great
payment, however, to find him delighted with everything,
and every one pleased with him. He goes on Monday, which
I think is a comfort, for he might by staying too long, leave
a different impression.

Last Monday we had engaged to go to Kinneil (Mr. Dugald
Stewart's) with Dr. Parr. We had two very agreeable days,
and returned here on Wednesday. I was rejoiced to be again
with my dear pets, who I had not left for a day this year and
a half. Nothing can be more entertaining than Dr. Parr has
been, thundering out all his political opinions, opening the
eyes and ears of our friends. He says he "*neither speaks
sluggishly nor thinks sluggishly.*"

From Dr. Parr.

October 27th, 1819.

DEAR AND EXCELLENT MR. HORNER,—I thank you for your
friendly and interesting letter. I sympathise with you in
your triumph over the Melville Vassals; I congratulate your
noble university upon the election of Mr. Wallace *. No
portion of my life ever was so happy as my late tour; no
part of that tour was so instructive, so delightful to me, as
the time which I spent among the hospitable, and most
enlightened inhabitants of Edinburgh. I shall ever look
back to their learning and their knowledge with admiration,
and to their hospitality with gratitude unfeigned. Your
message shall be delivered to my companions. I have
recommended the book which Mr. Pillans† put into my
hands, as worthy of attention and encouragement from the
teachers of our public schools, and I should suppose that he

----

* To the Chair of Mathematics vacated by Professor Playfair.
† Rector of the High School.

will receive several orders. I have reasons for requesting that you would send me an accurate description of Mr. Thomson,* and Mr. Playfair. Mr. Thomson is one of the most accomplished of all human beings, and Mr. Playfair is sweet-tempered, cheerful and very ingenious indeed. Mr. Horner, I cannot find language sufficiently strong to express the esteem, the regard, and the respect which I feel for you and your family. I shall venture to call you my friends, and I am not accustomed to employ that sacred word lightly. I have read Professor Brown's† book most attentively, and hereafter I shall say to him something about it. I am entirely with him, and you shall one day or other know my reasons. Remember me to all the Horners, Mr. and Mrs. Jeffrey, Mr. Playfair, Mr. and Mrs. Pillans, Mr. and Mrs. Morehead, Professor Brown, Dr. and Mrs. Gregory, Mr. and Miss Mackenzie, Mr. Alison and his household, and to Dugald and Mrs. Stewart. I do not forget the Principal Baird, nor Professor Dunbar nor Mr. Carson. I am overwhelmed with business, but you will soon hear from me again, I have the honour to be, with great respect and regard,

Your friend,

SAMUEL PARR.

From Dr. Parr.

November 5th, 1819.

DEAR, EXCELLENT, LEONARD HORNER,—To the latest hour of my life I shall remember with delight and with gratitude the hospitality, the good manners, the ingenuity, the taste, the learning and the science which distinguished the society into

---

* Thomas Thomson, Esq., Advocate.

† Dr. Thomas Brown successor to Dugald Stewart as Professor of Moral Philosophy.

which I had the honour to be admitted at Edinburgh. I
promised Dugald Stewart an engraving of myself, and a
metaphysical book written by a disciple of Kant. The book
and the engraving are ready, and they will be accompanied
by three other engravings, which you and Mr. Jeffrey, and
Mr. Thomson, will have the goodness to accept, as marks of
my deep and sincere regard. I am puzzled a little about the
best mode of conveying them, and at last I have resolved to
put them into a box, and I shall send it to you. You will be
kind enough to take care that the engraving and book be
conveyed to our respective friends, and with them you must
settle the price of carriage. We are to have a gag bill and
severe taxes. You see that the Prince Regent follows the
example of the German despots, and has issued a decree for
restraining the press at Hanover. Leonard, the Whigs must
be more united, more vigilant, courageous, and they must
make their hostility to Radicals and to the ministry more
conspicuous and efficacious. You will remember me in the
strongest terms of kindness and respect to all your name-sakes,
and to all my friends at Edinburgh. Tell Mr. Pillans that
there will be several orders from this country for the little
grammatical book. Leonard, you must either pay the postage
for your letter, or you may enclose it for Thomas Denman,
Esq., M.P., Russell Square. In a fortnight I shall write to
Dugald Stewart. I met Sir Walter Scott at the Archbishop
of York's Palace, and my scribe knows that the Tory had no
advantage over the Whig. I wish that Jeffrey and Thomson
had been of the party. Tell Dr. Thomson and Brother
Alison, that I shall never forget them. I am truly your
friend,

S. PARR.

From Dr. Marcet.

Geneva, *November 11th,* 1819.

MY DEAR HORNER,—I really feel ashamed at my long silence, especially as I was many weeks ago favoured with a most kind and interesting letter from you which I ought to have immediately acknowledged. There is a strong report here that the *Edinburgh Review* is going to be abandoned, I hope and trust it will not be so. There may have been lately some weak articles; but it is still, upon the whole, infinitely superior to any other, and the loss to the public, in instruction and rational amusement, would be truly irreparable.

I have to communicate to you a circumstance which I am sure will interest you, though in one sense it may give you some regret. We have bought an estate here of considerable value (eleven thousand pounds), and of great beauty; it is that in which we have spent the summer, and where it would afford us the most heartfelt pleasure to receive you and your family, if you could treat us with a visit next summer. The whole scenery of the Alps, presided by Mont Blanc, form an amphitheatre before us, and the lake and various other beautiful objects form our foreground. The house stands upon a rising bank. We thought it right, all circumstances well considered, to buy a *little bit* of land, and though this was rather larger than we wished (about three hundred acres), we could not resist the temptation. This, however, does not at all change our plans of future life. Since the restoration of Geneva, I always intended to divide my existence between the two countries, and Mrs. Marcet never felt any disinclination to this plan. But if, in order to possess an estate in Switzerland, it had been necessary to give up England, depend upon it, I never should have dreamt of acquiring one at that price. Our firm and positive intention is to put the whole of our land into the hands of a farmer, when we come back from

Italy, and to return to England towards the autumn of 1821,
at least. And we earnestly hope that we shall be able to
take our Scotch trip in the following summer. These periods
future contingencies may force us to alter. But I do not
think that any but very unexpected events could induce us to
alter these arrangements. The political events of England
keep us here in a state of great anxiety. I am afraid, after
all, this is less a political struggle than a war between the rich
and the poor. It has struck me more and more for some
years past, that however well the political liberty of the
English subject is secured, there is a kind of oppression
in England, the effects of which are most tremendous. I
mean that which arises from poverty, especially when
contrasted with the most colossal fortunes. There lies,
I think, the true danger. We shall be forced, in order to
defend our pockets, to side with my Lord Castlereagh, and
his regiments, which is a dreadful alternative; unless that,
by timely concessions, all the middling classes, who are as yet
but passive spectators, should be gained over to the cause of
the loyal and peaceful, so as to force by degrees the lowest
classes back again into their allegiance. This however, is I
fear, but a very improbable contingency. The early events
of the French Revolution, have rendered concessions very
unpopular in England, and although the two cases are far
from parallel, they are but too often assimilated by the timid,
as well as by the enemies of all improvement.

Mrs. Marcet begs to be remembered most kindly to Mrs.
Horner and yourself, Louisa and Sophy do not forget their
friend Mary, and beg also to be remembered. Louisa grows
a fine girl, and I hope will prove also a good one. Sophy is
also improving, and Frank is a good student, and a perfectly
happy and contented young man.

Yours ever most truly,

A. M.

To Dr. Marcet.

Edinburgh, *November* 16*th*, 1819.

My Dear Marcet,—I wrote to you some time ago, I think about two months since, and I hope soon to hear from yourself that you have received my letter, and to hear something of you and Mrs. Marcet, for except one short notice from Whishaw* two days ago, I have heard not a word of your proceedings since your arrival in Geneva. You know how interesting anything concerning you will be to us, and therefore if you have not written to me when this reaches you, do not delay it much longer. The rumours which I have heard forebode sad tidings, and I fear I must make up my mind to the sad conclusion that our future intercourse will be wholly in correspondence. The changes that the last few years have brought about to Mrs. Horner and me, have, with very few exceptions, been accompanied by sorrow only, and five years' residence here have not replaced the excellent friends I lost by quitting London.

I have at length got my wish accomplished, and your name will be spread in thirteen thousand different channels through the civilized world. The enclosed came to me this evening, wet from the press, and I do not wait for your having an opportunity of seeing the review when the book reaches Geneva, as you will be pleased to have an earlier sight of it. Dr. John Thomson has been overwhelmed with a work he has been printing all summer on the identity of chickenpox with smallpox in a modified state, and much to his regret felt obliged to say that he could not, at least for several months, undertake anything else. This postponement *sine die* I was very sorry for, as the review ought to have appeared more than a year ago, and I therefore resolved to

---

* John Whishaw, Esq., of the Chancery Bar; afterwards a Commissioner for auditing the Public Accounts.

give up the idea of a medical article, and to make it chemical
and statistical. I thought this would at least not make it
less adapted to the publication it was to appear in and
equally serve the object I had in view, to bring the book into
the notice of others besides the medical profession. I took
advantage of MacCulloch being here a short time ago, who
very kindly gave me his assistance, and with his notes and poor
Gordon's introductory part, I have made it out such as it is.
It is very incomplete without the medical part, but you know
how I have been baffled. I shall be anxious to know whether
you think my zeal has not done you more harm than good.

Whishaw tells me that you have bought an estate on the
banks of your noble lake; that is very natural, and will be a
source of great happiness to you, I have no doubt; but he
adds, that you are appointed a professor, and that you are
talked of as a candidate for a seat in one of the Councils; all
this, too, may be natural, but it looks so like a renunciation
of England that I grieve to hear it. Bur I rely a great deal
upon the ties which connect you with England, and upon
your experience of the high state of human existence in the
society of London in which you lived. Warm as your
feelings are for your native country, you have been too long
accustomed to a wider range; and you will, *I hope*, find that
your mind cannot contract itself within so narrow a space,
and that the true way to preserve a full relish for Geneva is
to re-visit it in some degree as a stranger. You have so
much better sources of political news than I can have in this
remote province, that I do not enter upon that subject. But
there never was a time when it was of deeper interest, for we
stand in a perilous situation, not from the Radical reformers,
but from those whom the Radical reformers have terrified,
and from the advantage which ministers will not fail to take
of these terrors to repress the progress of right opinions on
the relations between the governed and the Government, and

to encroach upon the liberties we now enjoy. There is unquestionably much misery among the lower orders in the manufacturing districts, but it is confined to these, and until that redundant population is removed, discontent will prevail.

I do not think I can give you any news from this place. Wallace of Sandhurst, the fittest candidate, has been appointed to the Mathematical Chair, vacant by the removal of Leslie to that of Playfair. Playfair has left no manuscripts in a state fit for publication, and very few of which any use can be made; not even an interleaved copy of the "Illustrations of the Huttonian Theory." A biographical memoir is anxiously wished for, but no one has yet been named for so difficult a task. Stewart is the only man alive capable of it, and when he is made fully aware of this, I hope he will undertake it. He is very well, and is, I believe, to print this winter the second part of his "Preliminary Dissertation to the Encyclopædia." Mary and my second daughter, Frances, are just now ill with scarlet fever, happily in a very mild form, and as yet no symptoms have appeared to excite any uneasiness, and as seven days have now passed I trust the worst is over. Mrs. Horner and the rest of my little flock are perfectly well.

I have only this little bit of paper left to send our very kind regards to Mrs. Marcet, and to our young friends, your children. Remember us also most kindly to those in Geneva to whom we are known, but most especially to Mrs. Schmidtmeyer, and every branch of the Achard family.

Believe me, always, my dear Marcet,
　　Faithfully and affectionately yours,
　　　　　　　　　　LEONARD HORNER.

From Dr. Marcet.

Geneva, 24*th December*, 1819.

My Dear Horner,—Since my last letter, various honours
have fallen on me, which I am afraid many people will be
disposed to laugh at. I have been elected a member of the
Representative Council, and soon after an honorary professor.
The former was a natural consequence of my having become
one of the great landowners of the Canton (three hundred
acres); but for the latter I am at a loss what excuse to
mention, unless it be that I offered my services to assist my
old friend, De la Rive, in a course of chemical lectures which
he intends to give this winter. Be it as it may, I shall take
care that all this does not turn my head, in which I hope I
shall be greatly assisted by the English ballast which is still
in it, and will operate as a useful counterpoise to all the
temptations which Geneva holds out to my vanity. I shall
not go over the same ground which I took in my last letter,
and which has not undergone the least change since I wrote
last. I cannot however, let your suggestion (*of visiting
Geneva in some degree as a stranger*) pass without entering
my protest against it. No, so long as Geneva shall continue
free, and its moral and political existence shall continue
respectable, I can only consider myself there, as in my own
native country; and I can never consider England in any
other light than my adoptive country. Whatever feelings of
love and gratitude I entertain for England, I cannot forget
that I can be but a mere spectator in its public affairs; and
although my taste or avocations might probably never have
induced me to take any direct concern in public affairs, the
absolute impossibility of doing it, must always be a bar to
perfect assimilation. This is a feeling from which I never
was able to divest myself, and the pleasure I now experience
in having a community of rights, ranks, and objects of

ambition, with those I associate with, however small the scale, is to me a very real enjoyment. Besides I really find a great charm in Republican forms of government. God forbid that I should ever recommend to do away with Royalties, peerages, etc., where they do exist, but I had much rather do without them, if an equal degree of civilization and refinement can be obtained without their assistance.

You talk of the *contracted scale*. True—very true—in comparison to London, Paris, Vienna, and perhaps one or two other first-rate cities—shall I put Edinburgh in the list of exceptions? but not in comparison to other cities, though perhaps ten times more populous. I find here *en petit*, all the elements of great and good feelings—public spirit in abundance—solid knowledge and great morality in the higher classes. I admit, however, that I could not remain here long without missing London exceedingly, but as I hope and trust I shall never *miss it long*, I enjoy my present happiness with as little regret or alloy as I can. I dare not speak of the present state of public affairs in England for fear I should be too despondent. England under a military Government, or under a rabble of Anarchists, what a dreadful option! and a kind of national bankruptcy in prospective. This state cannot last. There is, after all, too much sense and knowledge on board this vessel to let it sink. True patriotism will rise from its ashes. The spirit of Horner, and of Romilly, is not yet extinguished. Some great, but salutary changes will be made to restore to that beautiful structure its former energy and vigour. England will no longer be herself, if the excess of the evil does not soon suggest some appropriate remedy.

Ever most sincerely yours,

A. M.

CHAPTER VII.
1820-1821.

From Dr. Parr.

Hatton, *January* 17*th*, 1820.

DEAR AND EXCELLENT HORNER,—Last night I received your letter, it has relieved me from great anxiety about the box. Let me know when and where you will be in London, and whether I can send to you an engraving for Mr. Stewart.

I can scarcely hold a pen in consequence of an indolent humour of gout, which the cold weather has brought into my hands; write to me immediately, and present my best compliments to all your relatives, and to all the intellectual heroes and political worthies whom I have the honour to call my friends. Leonard, my Edinburgh visit was the most delightful, the most instructive, the most interesting period of my life.

I am truly your friend,

S. PARR.

P.S.—On Wednesday, the 26th, I come to my 73rd year. I shall have a sumptuous dinner and a most remarkable company. Oh! that you, Mr. Horner, and Dugald Stewart and Mr. Jeffrey were among us—drink to my health. Tell Mr. Jeffrey to inflict vigorous discipline upon Parson Philpot, of Durham. He is a very good scholar and a very vigorous writer, and a very crafty, ambitious priest, and a most intolerant and intolerable Inquisitor.

Remember me to Mr. Dugald, and Mrs. and Miss Stewart

Dr. Maltby desires his best respects to Mr. Stewart, and will soon thankfully acknowledge the letter he had the honour to receive from him.

Leonard! do not procrastinate.

[Mr. Horner at all times took the most anxious interest in
the education of his children, and he used to give his eldest
daughter, aged thirteen, lessons in Italian, before going out
on business. He was always an early riser, the breakfast
hour—winter and summer—being eight o'clock. He was
most orderly and punctual in all his habits. He ever had
the resource of literature and science in his leisure hours,
and kept up his interest in chemistry, but the study in which
he most delighted was geology. In the May number of the
Edinburgh Review he wrote an article on his friend, Dr.
Maculloch's book on the Western Islands of Scotland.]

From Dr. Maculloch.

June 2nd, 1820.

MY DEAR FRIEND,—Your review arrived yesterday, and I
need not say how much I am flattered by your treatment of
me. In the shape of praise, I must say I never saw any
thing better administered, and, upon the principle of *laudari
a laudates viro,* I shall begin to have a better opinion of myself
than ever I had before. Seriously, Playfair being dead and
gone, you were the only person left whom I thought capable
of understanding my views and appreciating my labours, and
as I was very much afraid that I had not done what I ought,
I was really anxious for your private approbation. I hope
still that I have *that* separately from what a certain sort of
etiquette makes it usual to say to the public, and if it be so,
I shall not probably do what I have long meditated, burn my
books like a repentant conjuror, and betake myself to a new
life. Some of these days, if ever I meditate more than a
mere technical work, I must ask your advice on the subject
of style. I am fully convinced that I have no talent in that
way, and my theory has always been that it was a matter of

feeling and soul, *nascitur non fit.* Yet a judicious friend
might point out much to be amended and thus improve that
which cannot be perfected. One may avoid errors and write
intelligibly, and do many more things, and yet be hard to
read. That the thing is unattainable by labour alone, is
proved by Gibbon and many others. I think that you would
write well, if you had a book to write—it certainly is
not a secret to be communicated. It must exist in the
sentiment, not in the diction only. When we meet perhaps
you will give me your opinions more fully than in the Review.
On your part of this subject I may say, in the first place, that
you have written a very agreeable and popular-looking article,
calculated to make geology go glibly down among the
readers of reviews. In the next place I think it calculated to
do a great deal of good by recalling the attention of our
namby pamby cockleologists and formation men to that which
is the proper business of geologists. Playfair, perhaps, would
not have done this better himself, and it will do you great
credit. It appears to me also that you have taken the right
line in this respect, as far as it related to the book. That
is the way in which I would have reviewed it myself, if I could
have done it. As to making any review of all the geology in
the book, that was plainly impossible, and I do not well see
what else you could have done, or how you could have taken
a better line than you have. Others would probably take up
other parts, and there is plenty left for any one who may
follow. Warburton has undertaken it for the *Quarterly
Review* at my suggestion; as Fitton deserted, after keeping
me five months on promises. What he may do I cannot
conjecture, but he will now have you for his model, and I do
not know that anyone would do it better than him. My
book on rocks is in the press, but unknown to anyone. I
have undertaken it on my own account, and must try, I
believe, to sell some copies by subscription. It will turn out

a thick, seven hundred page octavo. It consists now of the
introductory chapters, which you read, new modelled and
divided, according to your advice, with two additional elementary ones. The geological prefaces to each rock have been
reduced to mere matter of *apparent* facts, free of all speculation,
or of any matter which was not strictly rockological. Even
in this way, however, it turns out to contain all the
elementary matter of geology, strictly speaking, which
belongs to rocks, and is, *quod hoc*, an elementary work.
It was impossible to avoid this, because as my views (which
are yours, too) of a geological system have never been given
to the world complete by our predecessors, the catalogue of
rocks could not have been made intelligible without it.
Whenever that system shall become acknowledged, the
catalogues might stand on their own bottom.

Best regards to Mrs. Horner, and tell Mary to get plenty
of drawings ready against I come.

To Dr. Marcet.

Edinburgh, *8th November,* 1820.

MY DEAR MARCET,—It seems a very long time since a
letter has passed between us, but I daresay I am to blame.
I will not fill my paper with apologies, but with something
that you may care more to hear about. About a fortnight
ago, Mrs. Horner gave me a son; you know how much
reason I have to be pleased with my daughters, but I will
not disguise that I feel a little proud of this new acquisition
to my fortune. It is the more gratifying to me, because I
shall have the hope of prolonging the duration of a name
that must ever be dear to me, you need not be told that I
mean to name the little boy Francis. Notwithstanding all
this, however, I daily thank my stars that my other five
children are girls, for had they been boys, I should not have

known what to do with them. We have had a great deal of
anxiety lately, for last May and June we had our whole
family in the hooping cough, and our dear Mary was very
dangerously ill, for she had just recovered from the scarlet
fever. As change of air was strongly recommended, we
moved about for some time, and in July I took Mrs. Horner
and my two eldest girls to London. I ought not to say I, for
Mr. Lloyd, Mrs. Horner's brother, desired that they should
come, and supplied the means very liberally, of his sister and
nieces making the long journey with every degree of comfort.
We stayed there nearly six weeks, but a great many of our
friends were absent from town, and we often looked at Russell
Square with a sigh. We passed a day with Prévost and his
gentle, agreeable wife, but they, too, went out of town. Mr.
W. Haldimand I saw one day for a few minutes in the street.
I regretted very much that I had no opportunity of introducing Mrs. Horner to Mrs. Mallet, for she is a person I greatly
admire, but they left London for Geneva the very day of our
arrival in London. Blake was gone, and Wollaston had
mounted his short-tailed coat at Sir John Sebright's. The
only one of your friends of any eminence that I recollect seeing,
was that soft, delicate person, Pepys, with his shoe-brush
under his chin, as black and as bushy as ever. I had the
mortification of hearing some of the details of the new
Presidency of the R.S., and that if Wollaston had stood firm
he had a triumphant majority.

I got some details about you and Mrs. Marcet from the
Prévosts, and was very much grieved to hear that one of
your girls, Sophy, I think, had been very ill during a tour
you were making in Switzerland, and I conclude from what I
then heard, that you are all now on the Italian side of the
Alps, and that you mean to pass the winter in Italy. I trust
you will all keep your health, and if you do, I am sure you
and Mrs. Marcet will pass your time delightfully.

Mr. Whishaw paid us a visit in Scotland this last autumn, his great object was to see his ward, Mrs. Kennedy, in her own house, and was very highly gratified to see her in all respects so comfortably and so happily situated. Her husband is a most excellent man, very amiable and sensible, and he has attained a very respectable station for character in the House of Commons. It is something for Scotland to boast, to send so truly an independent member to Parliament. His place, Dalquharran Castle, is very nobly situated, and he is beloved and respected by all his neighbours. The day they arrived the whole population of an adjoining town had been assembled for some hours in the street through which they were to pass, and even the roofs of the houses were crowded; something of this was homage to the daughter of Sir Samuel Romilly, I have no doubt. William Romilly has been spending some months with his sister, and is now passing through Edinburgh on his return to England. From Ayrshire, Mr. Whishaw went to Mr. Stewart's, viz., to consult him about the publication of Sir Samuel Romilly's papers. The consultation was, as he told me, in every way satisfactory, and he came away with his mind quite made up as to the parts he should publish, having had the sanction of one on whose judgment and caution in a matter of this sort he placed more reliance than on that of any other person he knew. He afterwards spent three days in Edinburgh and I took him to Mr. Ferguson's at Raith, which he was very glad to have seen, and at the same time he heard Dr. Chalmers preach at Kirkaldy, and looked upon the house where Adam Smith was born, and where he wrote the "Wealth of Nations." You will, I am sure, be sorry to hear that Mr. Whishaw has finally relinquished the task he had given us some hope he would undertake, of writing a biographical memoir of my lamented brother. Ever since the sad death of Sir S Romilly he has never been able to

look at the papers, which had been put into his hands some
time before that event, and after having been repeatedly
disappointed of finding leisure, he saw such an accumulation
of employment before him, from his duty as executor of Sir
S. Romilly, that he could no longer in prudence defer coming
to the decision of giving up an undertaking he saw no
prospect of accomplishing within a reasonable period.

   After a great deal of anxious deliberation, it was at length
determined by him and by James Abercromby, that *I* ought
to undertake the memoir myself. You will not, I am sure,
suppose me capable of hearing such a proposal, without the
full extent of my incapacity rushing into my mind, but after
much meditation, which I would gladly enter into the
particulars of to you, if I could do so within the compass of a
letter, I resolved to attempt this arduous task. I would not
have entertained the idea of it for an instant if I had not felt
secure of such assistance, both here and in London, as would
secure me at least from doing injury to my brother's memory.
It is very satisfactory to me, that without any communication with Whishaw or Abercromby, the same proposal was
made to me by Sharp and Sir James Mackintosh. I have
thus before me, for the next two years, a task that will try
all my strength, and if I succeed so far as not unworthily to
hand down my name to posterity with that of my lamented
brother, it will be a greater honour than I ever expected to
attain. I am going in a few days to Kinneil, to spend some
days with Mr. Stewart in consultation upon this important
work. He is remarkably well just now, I have not seen him
for a long time looking more healthy or so robust. Mrs.
Stewart and her very agreeable daughter are also very well.
He has completed the second part of his " Preliminary
Dissertation to the Encyclopædia "; it is not yet published,
but it is printed, and I hear it is very good. I don't know
whether the last novels of Walter Scott had reached Geneva, the

" Monastery " and " The Abbot." They are greatly inferior
to what he had published before, and in England they are
considered to be decided failures. Notwithstanding all that,
however, he received three thousand five hundred guineas for
the said " Abbot," and he has *contracted* for two more, and
has received part payment in advance. One of them is to
appear next month, to be called " Kenilworth," and the scene
is to be laid at the castle of that name, near Warwick, during
the reign of Elizabeth.

I recollect no more literary news from this place. The
discovery ships, so long looked for, are arrived, having gone
five hundred miles up Lancaster Sound, where Ross stopped
short, and they were in one hundred and fifteen degrees W.
long., and seventy-five degrees N. lat. A north-west passage
is therefore probably practicable, but utterly useless from the
vast accumulation of ice. In science nobody here is doing
anything. Brewster is twaddling at polarized light, which
nobody understands or cares about ; and Tommy Hope is
gathering guineas—not laurels—getting up his chemical
drama in a new theatre, with great *éclat*. Maculloch is
going on with his " Geological Survey of Scotland." He has
lately published a geological book on the Western Islands,
which I have reviewed in the *Edinburgh Review*, and said
most conscientiously, that I consider it the most valuable
geological work that has appeared since the " Voyages de
Saussure."

We have no less than four Polish noblemen here, young men
of the first families, who are studying very hard. They are
very sensible and very amiable. There is the Prince Czartoriski, Count Sobieski, and two brothers—Counts Zamoiski.
They have been very much disappointed in not finding a
course of political economy delivered in the University, but
Maculloch, the reviewer of Ricardo's work, has been prevailed upon to give a private course to these Poles, and

a few more. To give right opinions on such a subject to young foreigners of influence in a part of Europe where they can be so little known, may be of immense effect. I shall advise them to translate Mrs. Marcet's book into the Polish language. I have never yet succeeded in getting justice done to that admirable book in the *Edinburgh Review,* but it does not stand in need of any support.

Should you visit Leghorn I trust you will not fail to go to the Protestant burying ground, where my poor brother lies, as I should be glad to have your and Mrs. Marcet's opinion of the monument that has been erected over his grave. It was partly designed by Sir Henry Englefield, and partly by Chantrey, and was executed by Micali of Leghorn, except the medallion which is, the work of Chantrey, and is, I think, very good.

If you go to Pisa, I hope you will make the acquaintance of Dr. Vacca, a most intelligent, agreeable man, and I believe the man of highest reputation in his profession in all Italy. If you see him, remember me to him in the kindest manner.

I will not enter upon the subject of the Queen; that odious business which will bring destruction on the country, I believe, for you will hear quite enough of it from other quarters; all I shall say is, that I was strongly impressed with a belief of her guilt before her trial, from the numerous reports one heard from every quarter, but there has been such an infamous disclosure of the most deliberate attempts at subordination, that I cannot attach any credit to the reports upon which my former conviction rested, and I am now very much inclined to believe her to be no worse than a coarse-minded woman, capable of saying very gross things, and forgetful of that reserve which belongs to her high station—offences that surely do not deserve to be visited with the punishment she has already received, far less that with which she has been threatened. In the meantime every other consideration is lost sight of and one would imagine that we had nothing else to attend to. The trade of the country is certainly in a very bad state, but it is in some degree better than it was, for the workmen are generally employed—there is, however, no advance of wages, and no increase of manufacturing profits, both of which are, I suppose, lower than they were ever known at any period. There is a most redundant supply of labour and of capital. The weavers' wages in the great manufacturing districts of Scotland, will not average 7s. per week, and the Banks, here and in Glasgow, not having the means of employing their capital on the spot, have been discounting bills in London, through the medium of agents there. Money can be had in London just now, for short periods, at two per cent., I am informed by an intelligent Edinburgh banker just come down, and yet the funds yield more than four; why? because the monied people, expecting some sad convulsion from this affair of the Queen, are watching for an opportunity of buying in at some very low price to which the stocks may fall in a panic. I need not say how happy I shall be to hear from you. Give our very affectionate regards to Mrs. Marcet and all your children that are with you. I should tell you that Mrs. Horner is making a good recovery, and that the little boy is stout and healthy.

Most affectionately yours,
Leonard Horner.

From Dr. Marcet.

Florence, *January* 5th, 1821.

My Dear Horner,—Your very kind and welcome letter reached me at Turin, where I spent a fortnight, and I avail myself of the first moment of leisure to answer it. . . . . Your biographical undertaking is certainly one which it required no common degree of energy and courage to enter

into. I have heard of your resolution with the greatest
interest, and with a pleasure not altogether unmixed with
apprehensions. I believe if there be a man who can speak,
even of a brother, without exaggeration or flattery, it is you
—but you will be perpetually labouring under the fear of
giving vent to your feelings, and it will be exceedingly difficult
to prevent your style from betraying your restraint. But as
your narrative will be principally made up of public events,
and of the delineation of a character which is generally
known and appreciated, and which I may truly say there can
hardly be two opinions, you will be, like every other author,
put upon your trial as a writer, but there is no risk of your
being severely judged as a historian. Pray inform me from
time to time of your progress. There never was, I can assure
you, a literary enterprise in which I felt more warmly
interested.*

We proceed on our journey without accident. We have
seen Turin, Genoa, Lucca, Leghorn, Pisa, and we are now
engaged in seeing Florence. We intend to remain here about
a month, and then go to Rome and Naples if we can. I am
much struck with the great and extraordinary objects of art I
meet at every step, but I confess that the political state of
this unfortunate country, too often takes off my attention
from inanimate objects.

The people seem everywhere to be awaking after a long and
painful slumber, though still struggling against the lethargic
influence of the old legitimate dynasties. Naples presents
the spectacle of a most curious phenomenon, viz., a nation,
which but a few months ago had no public spirit, no energy,
no knowledge of its rights; and which at the present moment
seems to possess all these qualities in a most remarkable
degree. Their moderation, too, is one of the most extraordinary features of the Neapolitan Revolution. Is it that

* Mr. Horner did not write the life of his brother Francis for more
than twenty years after this period.

liberty has really grown wiser by the example of other nations, or that the Austrian bayonets operate as the rod suspended in the middle of the school? Will there be a war or not? Will the king be firm, or will he be what kings too often are? I am sorry to say that various symptoms in his passage through Italy have given great doubts as to his sincerity, and have increased the probabilities of a war. What an awful moment for Italy. Many good Italians pray God that a war may take place, for they are confident that the whole of Italy will rise the moment that hostilities begin, and will expel its oppressors. They do not exactly know how to rally, for there is no rallying point; but they have a strong feeling that Italy requires an Italian Government, and that the different states must be cemented together by some bond or other. Naples is fully expected to take possession of the states of the Church, the moment that war is declared, and if its armies should obtain any small advantage at the beginning, there is no knowing what the consequences might be. Unluckily, however, the reverse is the most probable; and I own, for my part, that I had much rather see the warlike preparations subside.

England, poor England, is not much better at this moment than other countries! yet what superiority it has still over France in regard to the steadiness of its institutions, to the purity of its juries, and to the liberty of the press. How refreshing it is on the Continent, to meet with a *Times* or a *Morning Chronicle*. In science also and general literature, how superior your *Edinburgh Reviews*, or even your *Quarterly* (notwithstanding some infamous articles) to all other periodical works! I lament as much as any man, that England should not have better administration, but I hope to God that she will not run the chances of Revolution. There is too much to lose, and not enough to gain. The vessel must be patched up and refitted, for fear no sound materials could be found to

build a new one. Canning's retreat will afford some help to
the cause of moderate reformation.*

I long to hear how the Royal Society goes on with its new
President† and his valiant satellites. I do think that if you and
I had been in London, we might have had some little influence
in giving courage to Wollaston. Since he declined, however,
Davy was unquestionably the man; indeed, those who are
only acquainted with his scientific claims, think he was the
man at all events. If he assumes too high a tone it will be
our fault. In the meantime I enjoy the idea that science, in
England, will triumph over Princes, Dukes, and Chamberlains;
I hope you did not canvass too violently for your friend
Somerset.‡

Wollaston writes me word that Parry brought me eighteen
specimens of Polar waters. He has been applied to examine
them during my absence, and he says he has instituted himself
my *deputy*. You may imagine how flattered I feel.

I have arranged a nice laboratory at my country house of
Malagny. Berzelius compared it to the shop of a silversmith,
so new and bright everything is, yet I hope, however, that I
shall have time to tarnish some of my tools next summer,
though I fully intend to recross the Channel early in the
autumn.

Mrs. M. and I feel much interested in the birth and welfare
of a little *Francis* § in your family. Pray give me from time
to time fresh particulars, and remember us most kindly to
Mrs. Horner. My son is at Geneva prosecuting his studies

---

* In June, 1820, when the conduct of Queen Caroline was brought
before Parliament, Mr. Canning, who was then President of the Board
of Control, rather than bear any part in the proceedings, resigned his
office, and started on a tour on the Continent.

† Sir Humphrey Davy.

‡ Duke of Somerset.

§ His only son Francis Horner, born October, 1820, and who lived only
four years.

with uniform order and steadiness. He has not *tripped* once
in spite of all the frivolous English young men who abound
at Geneva. My Louisa is grown a stout girl, and draws
prettily. Sophy also thrives.

Yours ever affectionately,

A. M.

To His Daughter.

(Age 11 years.)

London, 21st *February*, 1821.

MY DEAREST MARY,—I have been dining with Mr. Hewlett,
and have come home early with the intention of going to Sir
Humphrey Davy's this being the day on which his house is
open to the weekly visits of such members of the Royal
Society, as will honour him by assembling at his house.
My last letter to your dear mamma would bring severe sorrow
with it both to you and to her. I am sure that few events
can have happened, perhaps none that would produce more
distress to you, than the death of your very worthy
and affectionate nurse, whom you have so long, and so
deservedly loved. In this, as in every other affliction, it is our
duty to bow with humble submission to the will of God.
You will always cherish a fond remembrance of her, and you
will have the comfort of thinking that you were always kind
to her, and that you contributed greatly to her happiness.
Her health has been getting worse, notwithstanding all the
care that was taken of her, and she has been carried off by a
disease which it was not in the power of medicine to remove.

Miss Edgeworth has just published two small volumes under
the title of "Rosamond" as a continuation of "Early
Lessons," which in the preface she says, is to be read only
by girls between ten and thirteen. I have got it for our
schoolroom, although I have not read it or heard anything
about it, but I think I can safely rely upon anything that

Miss Edgeworth writes for young people. I am glad you like
Hampton's "Polybius." He is the most authentic historian
of antiquity, and the translation is a very good one. Before
going on with it, I would advise you to read in Dryden's prose
works, a character or life of Polybius. I think it will make
you enjoy the history still more.

I look forward with great pleasure to our resuming our
Italian studies, I hope we shall have four months' steady
application without much interruption, and in that time we
should be able to do a great deal.

Farewell, my dearest Mary, and that you may always be
"the World's Wonder," is the fond wish of your very
affectionate father,

LEONARD HORNER.

From Dr. Marcet.

Florence, 5*th March*, 1821.

MY DEAR HORNER,—I wrote to you a few weeks ago, and
expressed my regret at having missed the opportunity of
visiting your brother's tomb. If it were necessary to show
that I was sincere in my regret, I could not produce a better
proof than the excursion I have just made from Pisa (where
I was induced to take a trip) to Leghorn for that purpose.
I can now tell you what I think of the monument. It is
truly touching by its good taste and simplicity. It would
have been so easy to adorn it with this or that symbol of
public services, or to swell the inscription with ostentatious
titles to public gratitude! The structure has sufficient mass
to catch the eye; but in the detail it can only be noticed by
the total absence of pretension or laboured embellishment—
by the eloquent simplicity of the inscription, and by the
great likeness of the *medallion.* My companion made me
remark that the form of the tomb might have been more elegant
if the base had been somewhat larger than the superstructure,

so as to resemble less a regular square, but I cannot say that
the circumstance would have struck me, or that I am quite
satisfied with the truth of the criticism. Mr. Martin of
Leghorn (Alex Prévost's brother-in-law) was so good as to
procure for me a little sketch of the tomb, which, however
slightly done, will serve to give you an idea of its appearance.
It is probable that you have already procured drawings of
this kind, but in that case, it may be acceptable to some of
the family, and at all events you will not scold me for having
ventured to inclose it in my letter. I have only to add on
this melancholy subject, that the Protestant burying ground
of Leghorn is one of the most soothing places of the kind I
have ever seen. The spot does by no means look forlorn,
and melancholy, as churchyards commonly do, and as it is
chiefly occupied by the remains of travellers from all parts of
the world, whom their health or curiosity, had brought to
Italy, the mind in walking through those monuments, is
rather agreeably hurried through a variety of recollections
totally unconnected with each other, an effect very little
resembling that which is produced by the monotonous legends
of a parish churchyard. Thus I very unexpectedly fell in
with Smollett's tomb; the next monument to one of Lord
Selkirk, whom I had known as Lord Daar in Edinburgh, &c.,
&c. In fact I left the place in a sort of pleasing reverie and
with a vague impression that one might lie there comfortably.

We are still here waiting to know whether or not it will be
practicable to go to Rome. We have now seen Florence
thoroughly well, and we have little to do but to see the same
objects over again, and to follow the stream of the Carnival
follies and dissipations; but a source of constant and prodigious interest to me, is the watching of the military and
political crisis which is taking place around us. The
Neapolitan cause, at first considered as desperate, inspires now
some degree of confidence, and hopes are at length entertained

(by those who feel an interest in the liberties of Europe) that
Naples will do her duty. If she does, it will be a noble
triumph ; for every difficulty, every unfavourable circumstance
seems to be attached to the Neapolitan cause, an infamous
king, whom a considerable portion of his subjects are still
inclined to obey, if they could do it without sacrificing entirely
their constitutional liberties, a most unequal war to be
carried on by a son against his father, by militias against
veterans, and above all by a people who have no military
renown to lose, and who, if they should feel cowardly in the
hour of danger, may always give to defection the colour of
loyalty. If the Neapolitans get over these difficulties they
will be great indeed.   In the meantime Italy awaits the event
with the most intense degree of interest, and reads her own fate
in the bulletins from Naples. As to Tuscany, it is peculiarly
constituted. The lower classes have no political feeling of
any kind, nor seemingly any susceptibility of excitement that
way ; while the higher, which are enlightened, and think
soundly enough on political subjects, have no energy or courage
whatever, and this state of inertia is farther strengthened by
the mild and excellent personal character and paternal administration of the present sovereign. The utmost length that
the boldness of a Tuscan can go, is to encourage the schools
among the poor. This they have lately done in a most
zealous and praiseworthy manner ; but a frown of the Minister
would suffice to put an end to them altogether, and that frown
is an event which may be hourly expected. They now say
that the old King of Naples is coming here to wait for the
event, and endeavour by his proclamations to excite defection
in the armies of his son. If you are Leonard Horner still, I
am sure your blood is not far from the boiling point at this
moment. There has not yet been any actual engagement.
But the Neapolitan columns reach Rieti, and the Austrians
are concentrating themselves at Spoleto. This looks like an

approach of hostilities. If Europe had any spunk she would immediately interfere. England, France, Spain, Portugal, Piedmont should form a congress and despatch a person of weight to arrest the progress of this atrocious invasion. Such a mission would be worthy of the conqueror of Waterloo, if he would act heartily in the cause. You see that I write pretty freely upon these subjects, though from a place overrun by Austrian bayonets. A Tuscan would set me down as insane if he saw my letter. Adieu, answer me at Florence *au soins de Mess. E. Fenzi and Cie.*, I may probably no longer be here at that time, but the letter will follow. You must not be quite so radical in your strictures when you write to me as I am myself, because times may become more arduous still. About the middle of May we shall repass the Alps to enjoy my native air again, my fine view of Mont Blanc, my new laboratory, my old friends, and above all, a truce to kings, queens, ministers, etc,

In the month of November I shall see good old England again—good, no doubt, though not so much so as it should be, but excellent in comparison to the 9999th part of the world.

Ever yours most affectionately,

A. M.

———

To Dr. Marcet.

Edinburgh, 10*th April*, 1821.

MY DEAR MARCET,—About three weeks ago I received your very kind letter, dated from Florence, the 5th of last month. Had it not been for some accidental interruptions, I should have sat down within a few days to thank you for all your letter contained. I cannot tell you how much I felt, how much we all felt, your affectionate kindness in the interesting account you sent to me of your visit to my poor brother's grave. It is the first report we have had from any one who

feels as you do, since the place where he lies was marked out
by the monument, and to have received so favourable a
testimony of the taste in which it has been executed, is at
once most gratifying and consolatory. I think the medallion
very good, and more pleasing than the bust—a cast of it is in
the room where I now write. I am very glad, too, that you
like the inscription, that in English was written by my father
and myself, and we asked his friend and school-fellow, Pillans,
to convey the same ideas in Latin. Your little sketch I value
very highly indeed, not only for the story it tells, but because
it will always remind me of your friendship and attachment.
I shall not forget to tell Alexander Prévost's pretty and
pleasing wife, when I next see her, that I have had reason to
know the kindness of her brother at Leghorn.

I cannot advance another step without mourning over the
fate of unhappy Italy, doomed still to suffer degradation and
oppression. She has shown, however, that the dawn of
liberty which is rising over Europe, has not yet risen above
the Alps, and the Italians of these degenerate days have yet
to learn that liberty must be bought at a great price. Notwithstanding this check, the day is not far distant when the
governments of Europe must assimilate themselves with the
progressive improvement of their people, or be overthrown.

You will be glad to see that the Catholic Question has at
last passed the House of Commons, after a more strongly
contested battle than has ever taken place, for there were at
least six great divisions upon it. I have no expectations that
it will pass the House of Lords at present, nor at any time
until the Court are for it. But already an important step
is gained, and the ultimate success is infinitely nearer than it
was, before this decision in the Commons took place.
Another great question is making a decided progress, the
very important one of Parliamentary Reform—which is the
only good result that has arisen from the odious and

disgusting trial of the Queen, for the division on that
question, shewed to the most obstinate unbeliever, in colours
which he could not fail to recognize, that public opinion has
not the slightest effect upon that house, and this belief has
been brought home to some of the most considerable opponents
of any change in our representative system, as may be seen
by the late meetings for that express purpose in the counties
of Suffolk and Huntingdon.   But keenly as we both feel upon
political questions, I cannot indulge in free discussion with
you at present, and will therefore drop the subject.

I made a hurried journey to London in February, but can
give you very little information about our common friends,
because I was only a fortnight there, and with my mind
wholly occupied upon business.   I did not fail, however, to go
to the Royal Society, having rather a curiosity to see how
Davy would *look* the President.   While I was waiting in the
ante-room, talking with Allen and Dick Phillips, I saw, as I
thought, one of Sir Humphrey's footmen come bustling into
the room, a short time however discovered that it was the
Knight himself, attired in a court dress, swaggering up to the
chair.   He had got on the coat and waistcoat of the court,
but without either lace to his shirt, or powder on his hair, and
when he put on a very grotesque cocked hat with a cut steel
button, he looked very like the porter at a shabby nobleman's
gate towards the close of his livery suit.   This imitation of
Sir Joseph is mighty ridiculous, and the more especially
as Davy is the most awkward man in the world, and
accordingly he sometimes puts on his cocked hat wrong side
foremost, as I am credibly informed, while he is sitting in the
chair.   I went to his evening party, which I did not find very
agreeable, from the inconvenient room it is held in, but he
was very accessible and civil.   I am happy to hear that he is
working steadily in his Laboratory for several hours a day, but
upon what subject I did not distinctly learn—some one told

me the magnetism of galvanism.   If he works for several hours
a day, he is sure to do some good.   He complains bitterly of the
encroachment upon his time, by the constant applications to
him for advice as President of the R.S.   I saw Wollaston for
a very short time, for I was prevented from dining with
him at Blake's; I never saw him in better health.   I have
been very unfortunate in not seeing Mr. W. Haldimand for a
long time; for every time I have been in London, I think for
the last three visits, he has been in Paris.   He does not appear
to take so active a share in the business of Parliament as I
thought he would have done, especially on questions of finance
and trade.   I went to the Geological Society, which seems to
me to have got into very feeble hands and to want a great
deal of the energy it had in former days.   I fear that Warburton, Blake, and Wollaston go very seldom near it.
Greenough is now in Italy I fancy, or in Greece.   I should
fear that his health is not equal to the labour of geological
investigations in the hot climates of the East.   .   .   .   .   ..
You will be sorry to hear that Stewart has been very seriously
ill, but he is better.   I fear, however, that he must give up
every pursuit that requires any intensity of application, and
that he will not complete his great work on the Philosophy of
the Human Mind.   He has finished his Preliminary Dissertation to the Encyclopædia.   He has promised me some
assistance in the work in which I am engaged, and I trust he
will be able to fulfil that promise.   I never cease to abuse
MacCulloch for having failed in his promise of reviewing the
" Conversations on Political Economy."   He dined with me
last week, and he at last fairly acknowledged that he had
begun a review, and found he could not do the book justice.
He added, however, before a pretty numerous party, that " it
is the best book on the subject without any exception," these
were his words.   I am now engaged in carrying into effect a
scheme I have had for some time in contemplation, of estab-

lishing a school for the instruction of the mechanics of
Edinburgh in such branches of science as will be of practical
benefit to them in their several trades, I wish to put that
information within the reach of their fortunes, and my plan
is so far matured, that I expect to have a course of lectures
delivered next winter on practical chemistry, and mechanical
philosophy, and to lay the foundation of a library of books on
popular science, which shall circulate among the artizans.
They are to pay seven shillings a quarter, and if the public
of Edinburgh will give one hundred pounds a year, there will
be no fear of success. If you would like to hear further
details of this scheme, I shall be very glad to send them.

Mrs. Horner and all my children are quite well—my little
boy thrives, and promises to be stout and healthy, and Mary
is in better health than she has been for a long time, and
besides growing tall is growing in breadth. Remember us all
very affectionately to Mrs. Marcet—how happy it would make
us to see you here. Do not allow Louisa and Sophia to
forget Mary, for they may chance to meet, and I should be
sorry that it should be as strangers. Remember me very
kindly to Frank, who has now grown out of my remembrance
I daresay. My youngest sister was lately married to Major
W. Power of the 7th Dragoon Guards, and we have every
reason to be pleased with the marriage. She is now at
Nottingham, where the regiment is. Mrs. Kennedy has been
in Edinburgh, and has gone back to Ayrshire, she is a
charming person, is very popular indeed, and seems perfectly
happy. Indeed I never saw two people more devoted to each
other.

Pray write to me soon again, for your letters are a very
great treat. God bless you.

Affectionately yours,

L. H.

London, 11*th August*, 1821.

My Dear Marcet,—This will be delivered to you by my
very good friend Professor Pillans of Edinburgh. He is on his
way to Switzerland with his wife, and a young Polish
nobleman, the Prince Czartoriski, who has been for some
time under his charge. They are all worthy of your attention,
and fully capable to understand and to relish everything that
you believe will be interesting to cultivated and intelligent
minds. Any kindness that you can show them I shall be
very grateful for. They know and live with all the best
people at Edinburgh and know your friends well.

I have been in London about a week, but find very few
people in town. I have seen Mr. Haldimand and got some
account of you and Mrs. Marcet from him. He holds out a
very agreeable prospect of your being in London next winter.
If you come I trust I shall certainly see you, and perhaps
Mrs. Horner may too. I shall not say how very great the
pleasure would be to us both, and you will understand that I
use *you* in the plural sense. I left Mrs. Horner and all my
children quite well—my boy growing a stout, healthy child.
I hope all yours are as well. Remember me with great regard
to Mrs. Marcet.

I am always, very affectionately yours,

LEONARD HORNER.

[Mr. Horner was much occupied this year in establishing
the School of Arts, a college for working men in Edinburgh.
He was calling one day at the shop of Mr. Bryson, an
intelligent watch-maker, when in course of conversation he
asked him whether young men brought up to the trade of
watch-making, received any mathematical education; Mr.
Bryson replied, that it was much to be regretted that, owing
to the expense and usual hours, at which mathematics was

taught, it was quite out of the power of working mechanics to receive such an education. From that moment Leonard Horner set to work to supply this deficiency. The result was, the School of Arts was established, under the direction of the following men :

Dr. Brewster, afterwards Sir David Brewster.

Professor Pillans.

John A. Murray, afterwards Lord Murray.

George Forbes (Treasurer).

Leonard Horner (Secretary).

Jones Skene, of Rubislaw.

James Jardine, Civil Engineer.

The Deacon of the Free Corporation of Hammermen.

The Deacon of the Corporation of Mary's Chapel.

Mr. James Milne, Brass Founder.

Mr. Robert Bryson, Watch Maker.

Mr. John Ruthven, Engineer and Printer.

The Prospectus for this college bore the title of " School of Arts for the better Education of the Mechanics of Edinburgh, in such branches of physical science as are of practical advantage in their several trades."

It states that "however eager a young watch-maker or dyer may be to understand the theory and principles of the art which forms his daily occupation, and by which he lives, he has no means of obtaining that information because the expense is greater than he can afford. That the most intelligent master mechanics in Edinburgh all agree in lamenting the imperfect education of their workmen."

The School of Arts was opened on the 16th October, 1821, under the presidency of the Lord Provost. The two leading classes then established, were Chemistry and Mechanical or Natural Philosophy.

Mr. Horner delivered an introductory address at this meeting.

Henry (Lord Cockburn) writes in the Memorials of his time (a posthumous work published in 1856), as follows:—

"It was in October, 1821, that an institution for the instruction of mechanics, since known as 'The School of Arts,' was opened in Edinburgh. If not the first, it was certainly the second establishment of the kind in Britain. The whole merit, both of its conception and of its first three or four years' management, is due exclusively to Leonard Horner. His good sense, mildness, and purity made it a favourite with the reasonable of all parties and classes. It has gone on prosperously ever since."]

From Henry Cockburn to Thomas Kennedy.

14, Charlotte Square, Edinburgh, 25*th October*, 1821.

HORNER's School of Arts has been opened gloriously. He is a most useful citizen, and it is of great importance to have such a person here, not of the law. I have no doubt that with his excellent habits of arrangement and of business, of good manners, science and whiggism, he will in time greatly raise the character and zeal of our merchants and tradesmen.

Extracts from letters of Mrs. L. Horner to her Brother.

Great King Street, Edinburgh, 23*rd October*, 1821.

IT is now a fortnight since you left us. Leonard sent you the paper on Saturday giving an account of the School of Arts, now amounting to three hundred and thirty students. To the first lecture each student was permitted to bring two friends, and consequently above six hundred were admitted and forty were sent away. We have heard both from letters and verbally, many flattering things said of Leonard's opening address. The number of students increase daily; three lectures are commenced, one on mechanics, one on chemistry, and one on architecture, a fourth on farriery will begin in a fortnight. This is very promising.

All the children are well, we miss you and find now many things that we ought to have shown you.

October 29*th*,—We have delightful weather just now, and only want you to walk to Arthur's Seat with us, which from the bad weather, we did not accomplish. Mary has learnt one hundred and fifty lines of Tasso, which she can repeat with great ease.

The School of Arts has arrived at four hundred; no more can be admitted, though many have applied. The reading-room and lectures go on admirably, and I hope in a few weeks that Leonard may leave it to flourish by itself.

November 22*nd*, 1821,—The School of Arts flourishes much under Leonard's care and the lectures give great pleasure. The number of books given out on the last delivering day (which happens once a fortnight), was two hundred and seventy, and from thirty to fifty men sit in the reading-room every evening to read, which they do with silence and attention. One of Mr. Trotter's (the upholsterer), joiners, has consequently proposed to teach a number of men mathematics and geometry, and he has taken a house on the Castle Hill, and has got thirty scholars, some of those who go to the School of Arts.

———

Edinburgh, *November* 22*nd*, 1821.

MY DEAR MARCET,

I suppose you are by this time once more in London, to the great comfort and satisfaction of many who will be delighted to have you and Mrs. Marcet again amongst them, and I trust not without a good deal of satisfaction to yourself. Your letter of the 23rd October gave me great pleasure, and particularly that part of it which holds out a prospect of my

seeing you here. I need not say how glad I shall be to see
you, and my young friend Frank, and should you determine
to place him here, I hope you and Mrs. Marcet will have the
comfort of feeling assured that he has friends who feel a warm
interest in him. If there are any points of preliminary
information which you wish me to obtain for you, you know
you can freely command me. You will come at a favourable
time for seeking all those you will care most to see, as people
are now all collected for their winter occupations. You will
find the University in a very degraded state from that in which
you left it. Black, Stewart, Playfair, Gregory, Monro, are all
replaced by very unworthy successors.

My friend Professor Pillans, has told me of your beautiful
place, and of his having seen Mrs. Marcet, but he was much
disappointed that he did not see you, and that he was obliged
to make his stay in Geneva so short. He is a most excellent
man, and one of the few props of our University, but unfortunately not in one of those chairs, whose celebrity extends
beyond the limits of Scotland. Mr. MacCulloch, whose merits
as a Political Economist are not unknown to you and Mrs.
Marcet, has just begun a private course of Political Economy,
preparatory to a public course next year. I heard his introductory lecture two days ago, which was very admirable, and
in recommending the books for his pupils to study, he
particularly noticed "the admirable production of Mrs.
Marcet."

I have been very much engaged of late in the establishment
of a new Institution here, the School of Arts for the better
education of mechanics. The success has been beyond my
most sanguine expectations, for the mechanics have come
forward so eagerly, that we have been obliged to stop the
further issue of tickets. We have 420 students. We have
four lectures a week, one on chemistry, another on mechanics,
on architecture, and on farriery. We have at this day a

library of nearly 500 volumes, which circulate among the
mechanics, and the students themselves have just established
a class for the elements of geometry, having found a young
joiner, who is a very good mathematician, who has undertaken
to teach the class *gratis*. I enclose to Mr. Haldimand some
printed papers giving some account of this institution.

Mrs. Horner desires her very kind regards to you and Mrs.
Marcet. We are all well, you will find your friend Mary much
grown, with four sisters and a brother. My father's family
are all well, and I am sure will be most happy to renew their
acquaintance after an interval of more than eight years. You
have said nothing about Louisa and Sophy, I trust they are
well.

> Believe me, my dear Marcet,
> Always affectionately yours,
> LEONARD HORNER.

1822.

[On the 24th April, Mr. Leonard Horner attended the close
of the lectures at the School of Arts for the season, and he
gave a farewell address to 400 students. About fifty gentlemen
besides were present. The students presented him on this
occasion with a beautiful silver inkstand with the following
inscription engraved on it :—" Presented by the students who
attended the first session of the School of Arts, to Leonard
Horner, Esq., as an expression of their gratitude to him for
promoting the establishment of that Institution. 24th April,
1822."]

Edinburgh, *May* 13*th*, 1822.

MY DEAR MARCET,

You have possibly heard from Frank, of my good intentions
of writing to you, which have not been fulfilled as I wished,
or as they ought to have been. I am glad to say that he is

very well, and I daresay before you come every mark of the
small pox will be effaced, at least such as are not indelible,
for there are a few places where the perfect continuity of the
skin is interrupted, when from modesty or any other cause,
there is a determination of blood to the skin of the face, there
is then a distinct evidence of what he has gone through; but
in the ordinary state of his complexion very little is visible.
He has, himself, told you how far he has been pleased with
his masters, MacCulloch and Galloway have both told me
that as a pupil they have found him quick, intelligent and
very zealous in his studies. He talks with great pleasure of
beginning a course of practical chemistry under Dr. Fyfe. I
hope you will have no reason to regret the step you took in
placing him here. I see no change in his manner from what
it was when you left him. . . . . I sent you a newspaper
some time ago, which gave you an account of the conclusion
of the first session of the School of Arts. I am now preparing
a report, which will give a detailed account of all our proceedings from the first establishment, and will serve, in some
measure, as a guide to those who are disposed to make the
experiment in other places. You may suppose that the notice
which the mechanics took of my services in their behalf was
very agreeable to me. I am busily engaged in organizing a
class for drawing to occupy the summer months.

Frank tells me that you saw Jeffrey and Murray, for he
gave me an account of there being a strong party of *cognoscenti*
in Harley Street. I have not seen either yet since their
return, if indeed Murray has returned. I hope you have
settled with Jeffrey about both reviews.

We are looking forward with great pleasure to your coming
here, and I hope you will not fail to arrive before the 1st of
July, that those whom you would like most to see, may not
be yet dispersed for the summer. It is very difficult to quit
so many attractions as London possesses, but I should imagine

that by the end of May you will be pretty nearly satisfied with the rich meats and sauces of London, and will be longing for a little of the game of the second course in Scotland, to say nothing of the *entremets* at Foston in Yorkshire.*

Mrs. Horner and Mary are both well, and look back upon their late visit to London with great pleasure, as having afforded them an opportunity of keeping up many friendships of great value. I was very happy indeed that Mary had it in her power to become better acquainted with Louisa and Sophy. When they come here they will be able to say some words in Italian to each other, for Mary is working hard under a very good master, one very well known to Mrs. Somerville in Italy, Signor Bugni, the translator of Mrs. Barbauld's "Hymns for Children," dedicated to Mrs. S., which you have no doubt seen, if by chance you have not, get the book, for you and Mrs. Marcet will read it with much pleasure. Remember us most kindly to the Somervilles, to Warburton, Wollaston, Mr. Haldimand, and especially to the Mallets, two of the most agreeable people I know.

Believe me, always, most faithfully yours,

LEONARD HORNER.

From Dr. Marcet.

Culbokie (Mr. Fraser's), about 30 miles North West of Inverness,
25th August, 1822.

MY DEAR HORNER,—Here we are, really lost in the wildest parts of the Highlands, leading a most singular wandering and *parasitical* life, which is not without its charms. We had letters for no less than four Frasers, all of whom we have visited, and several more besides. We first began with the Macphersons (Dr. Brewster's relations) who are truly very agreeable people. They have (to use Davy's old phrase) *a very good tone of mind*, for their house is remarkably com-

° The Rev. Sydney Smith's Vicarage.

fortable, and their hills covered with birds, but, independently
of these important accomplishments, they are really a very
pleasant family. Mr. Fraser of Belladrum is another very
pleasant house in the middle of beautiful Highland scenery,
for which Mr. Cranstoun had procured us letters. There we
dined, and I shot one morning in company with Sir Humphrey
Davy, who having an idle day at Inverness condescended to
come and see us at Belladrum. We killed a black cock each,
and both crowed famously upon the glorious occasion.

From this place Mrs. Marcet with her children and servants,
set out this morning for Inverness and Fort Augustus, at the
last of which places they will arrive to-morrow evening, and
as to myself, I was left behind for the purpose of walking to
Fort Augustus to-morrow, across the hills, shooting all the
way (fifteen miles as the crows fly) so that while they perform
a journey of sixty miles in two days by the road, I rest one
day, and have only fifteen miles to walk the next, through
the purple hills. This, you will allow, is at least a curious
life, and certainly not very similar to that which I have led
in London for 25 years.

All the Highlanders I have hitherto met (with the exception
of Grant of Rothie-Murcus) have been Tories, but I must say
that their tone is so moderate, and their opinions so bordering
upon Liberalism, that I have found their conversation easy
and inoffensive, and have had, hitherto, no occasion to wish
that they were otherwise, for a shade of difference in political
opinions is more favourable to conversation than a perfect
coincidence, from which, if any conversation arises, it is often
merely for the purpose of indulging in abuse and exaggeration.

Mrs. Marcet received, a few days ago, a letter from Mrs.
Horner announcing your return and enclosing some valuable
letters. Roget wrote me word, that you would perhaps bring
with you some copies of my paper on Sea-water, but I suppose
it was not ready when you left London. The mountains are

very fine; but plenty of newspapers and of friends to converse with, are a blessing without which the mountains soon lose their charm. Pray, on the receipt of this, write to me Post Office, Glasgow, where we expect to be on the 1st September, and tell Mr. Jeffrey that we shall be at Dr. Robertson's on the 2nd at the latest, when we hope and trust we shall have the pleasure to see him. You should ask Mrs. Horner's leave to come and pay us a visit there, or rather you should bring her with you to see the falls of the Clyde, and take a trip to Loch Lomond.

   P.S.—Fort Augustus, Monday night. We are just arrived here from our respective expeditions, and I brought eleven grouse in my bag.

Ever yours most truly,

A. M.

Edinburgh, *September 14th,* 1822.

My Dear Marcet,—Wollaston arrived here on the 11th, he is with the Somervilles, and they go away in a day or two on a little trip to the Highlands; the Somervilles return to Edinburgh, but Wollaston returns South by the West Coast.

We have been much pleased to hear that your Highland journey was so successful, both in regard to your particular enjoyments, and that Mrs. Marcet and the rest of the party did not find the country uninteresting from its being a humble similarity to Switzerland, but that it possesses merits peculiar to itself. I am very glad, too, that you found the people so agreeable, and entirely agree with you, that it is a great mistake to suppose that difference of political opinions is unfavourable to a pleasant intercourse with strangers, provided there is a fair tone, and a disposition on both sides to give due credit to the sincerity of each other. Very soon after the 12th of August we received a box with three brace of moor fowl, which we consumed in the faith that they were your shooting. I hope you did not leave the Highlands without killing a red deer.

Since my return I have been engaged very much about the School of Arts, and if you have seen any of the Edinburgh newspapers of last week, you will have seen that B—'s iniquities have been fully exposed, and that he has met with a complete defeat. The School is now more firmly established than ever, with an excellent set of directors, and every prospect of its going on better than ever.

If you have not seen the account of the proceedings let me know, and I will send you a copy of a newspaper containing it.

Give our kindest regards to Mrs. Marcet, your son, and the young ladies.

Ever, my dear Marcet, most truly yours,

LEONARD HORNER.

From Dr. Marcet.

Carlisle, *September 15th,* 1822.

MY DEAR HORNER,—Our Scotch tour is over, and I wish I could have taken a final leave of you, and expressed to you my sincere thankfulness for all your kindness to us during this interesting episode of our life, and for all the friendly advice and assistance you have given to my son during his stay in Edinburgh. We have accomplished all that we had intended, without the least accident, or untoward occurrence. We have paid our visits to Dr. Robertson, to Mr. Cranstoun, to Mrs. Kennedy, to Lady Selkirk, and have been received everywhere in the most cordial manner. It is impossible to leave a country with more agreeable recollections and associations than we do on the present occasion. We now intend to pass rapidly through England, stopping only a couple of days at Manchester, a couple of days at Birmingham, and a couple of days at Blake's in Hertfordshire. We shall, therefore, in all probability reach London about the 25th of this month. Mrs. Marcet is not yet tired of travelling. She would stop at every bush if I did not spur the party on. The girls are just as bad, and all would be ready to turn back, and pay another visit to Mrs. Horner and to Mary, if our destinies did not require us to turn our steps southwards.

Believe me ever, dear Horner,

Most faithfully yours,

ALEX. MARCET.

From J. L. Mallet, Esq.

London, *20th October*, 1822.

MY DEAR SIR,—I have undertaken the melancholy task of communicating to you the death of your old and valuable friend Dr. Marcet, and I can judge from my own feelings how deeply you will be affected by this unexpected and distressing intelligence. You may probably have heard that in the night of Thursday, 10th, or rather early on the Friday morning, Marcet, who was then at Westcombe Park, near Greenwich, with his brother-in-law, Mr. Morris, was attacked with a sudden and violent pain in the chest, shooting like electric shots to all parts of his frame, and which he immediately conceived to arise from gout in the stomach. He considered the danger as imminent, and as it appears, despaired of his situation. Frank hastened to town however, and in the course of less than an hour and a half, Dr. Roget arrived, and soon afterwards Dr. Babington. Brandy and laudanum were administered with success; the pain was relieved, and all immediate danger subsided. Nevertheless Marcet remained feverish, weak, and irritable, and Roget slept every night at Westcombe Park, from the Saturday to Wed-

nesday, on which latter day it was understood that the Marcets
might return to their lodgings in Great Coram Street, if the
weather should prove favourable. It proved rainy, cold and
boisterous, but our friend insisted upon returning, his mind
being intent upon setting off for Geneva as soon as
circumstances would admit. It does not however appear
that he suffered from the removal; but he continued feverish
and unwell. He was restless on Friday night; and remained
in bed on Saturday till three o'clock. He nevertheless
conversed for some time with Roget, when he called in the
morning, wrote several notes, ordered some fish for his
dinner and rose at three o'clock. When about half dressed,
he complained of faintness, sat down on a sofa and seemed
to faint away. Mrs. Marcet, who was in the room, immediately
sent for Dr. Roget. At that moment I accidentally came in,
and Dr. Roget was with us in less than five minutes; but
when he arrived, all pulsation was gone, the extremities were
cold, and Marcet had sunk apparently without a struggle;
Dr. Bright followed soon after, and all was tried that skill
and friendly anxiety could suggest, to restore the circulation,
but tried in vain. Mrs. Marcet was alone in the house with
her daughters, and the situation was extremely distressing;
for a long time she could hardly believe the reality of the
event: it seemed like a frightful dream, and her nerves were
greatly agitated; but when Frank returned home, she gave
way to natural emotions and felt relieved. Poor. Frank
himself was in a state hardly to be described; having left his
father at two o'clock well to all appearance. Prévost, who
was going into the country, received my summons in time to
return. The Morris's and William Haldimand, to whom
expresses were despatched in the country, likewise came at
night, and Mrs. Marcet was then sufficiently recovered from
the first shock, as to find some comfort in their affectionate
society and sympathy. She slept during part of the night,

and is upon the whole better to-day than I had expected.   . . .
I saw Marcet for the last time on Thursday evening. His
countenance was good and step firm; but his voice weak
and faltering, and his whole frame in a feverish
and irritable state. He related to me at some length the
circumstances of his former attack, and expressed satisfaction
that his mind had been so serene during the paroxysm, and
perfectly resigned to what he then considered as inevitable.
He did not think that he had to bear a further attack; but
observed that he no longer held life by the same tenure.
Alluding to the state of debility he then was in, he contrasted
it with his strength and activity during his tour in Scotland
spoke with delight of his sporting excursion, and said that he
never could come to England again, without contemplating a
visit to his friends in the Highlands. I left him, I will
not say without uneasiness, but certainly without any
apprehension, and yet it was the last time I was to press the
hand of so true a friend. His warmth of heart and social,
manly feelings, his enlarged benevolence and independence of
character, can never be forgotten by any of us. His loss will
materially alter the situation of his family; what Mrs. Marcet
will do ultimately, I do not know, but at all events she will
probably spend the winter in England. I will not close this letter
till to-morrow, and will then give you the latest intelligence of
your disconsolate friends. *Monday, five o'clock.*—I have just
seen Mrs. Marcet, who is composed and resigned, and better
in all respects than I could have anticipated. She accedes
to every reasonable suggestion, and is all that can be wished
in her situation. Her intention is to follow up her husband's
wishes, and to proceed to Geneva soon, in order that her son
may follow up his studies there. Perhaps you will have the
goodness to write a few lines to Lady Selkirk and Mr. Dugald
Stewart, who would be much shocked on seeing the
intelligence in the papers. Whishaw writes to Mrs. Kennedy

and would have written to you if I had not. He is quite
well. Mrs. Mallet is tolerably well and joins me in affectionate
remembrance to Mrs. Horner.

Ever, my dear Sir,
Yours most faithfully,
J. L. MALLET.

From Mrs. Marcet.

Grosvenor Street, November 18th, 1822.

I cannot quit England, my dear Mrs. Horner, without
sending a few words to you and Mr. Horner, whom
I love and esteem as one of my dear husband's most valued
friends ; I thank you both for your kind sympathy, and can
assure you that I am more calm and more resigned to this
awful dispensation of Providence than I could have thought
possible ; the retirement of my brother's country seat and the
country air have in a great measure restored my health and
strength, yet my misery is very great, and I am compelled to
look very often on my children to believe that I am not
wholly wretched ; if their tenderness and the warm feelings of
my friends could afford me consolation I should be comforted ;
but, oh, my dear friend, the whole world cannot replace a
husband—may Heaven long preserve you yours.

We are going for two years to Geneva, according to the
plan traced out by my husband, I can follow no other steps.
Frank is going to study the law there, and we shall not be
separated. Adieu then, my dear Mrs. Horner, remember me
affectionately to Mr. Horner, and my poor children unite with
me in love to Mary and the little ones.

Ever truly and affectionately yours,
J. MARCET.

To Sir James Mackintosh.

Edinburgh, *December 9th,* 1822.

MY DEAR SIR JAMES,—You are probably aware of the great influence upon the political feelings of this place, of the public dinners that have been held here, first on the occasion of Lord Erskine's visit to Scotland three years ago, and the two last anniversaries of Mr. Fox's birthday. The improvement in independence, and in the spirit to come forward to shew that independence is hardly to be believed. It is, therefore, of great importance that these public occasions of bringing the people together here should be managed with great care, and with that cautious preparation which is most calculated to give them weight and respectability. The Committee to whom it was entrusted at the last anniversary to make the preparatory arrangement for the next celebration of Mr. Fox's birthday, with the concurrence of those whose opinion is entitled to most weight, are most anxious that *you* should take the chair; they are satisfied that by doing so, you would give great importance and character to the meeting, and therefore most essentially contribute to its usefulness. It would preserve it in the true *Whig* spirit; and take away from the enemy the possibility of using their favourite cant of its being a *Radical* meeting. It is proposed to apply to Mr. Cranstoun to be your croupier.

Before making this application to you, it has been very deliberately considered whether there would be any injury to the good cause, considering that you come to Scotland to be installed as Lord Rector of a University, and it does not appear to us that it can possibly do so in any degree, this is the opinion of those who were most rejoiced when they heard that the political dinner of Glasgow was given up. Cockburn is entirely of this opinion, and we had some conversation with Jeffrey yesterday, who promised to write to you immediately, and I hope he has done so by this post.

o

The birthday of Mr. Fox is the 24th January, but as we understand that there is no chance of your being in Edinburgh at that time, the committee leave you to fix the day. Two years ago that meeting was held on the 12th of January. As it is of consequence to give as early notice as possible of the dinner, I beg the favour of you to let me know as early as you can with convenience, whether we are to have the good fortune to have your assistance in forwarding the independence of Edinburgh.

I am, my dear Sir, most faithfully yours,

LEONARD HORNER.

P.S.—I may mention that Mr. Allen concurs with us.

From Sir James Mackintosh.

Mardocks, 15*th December*, 1822.

MY DEAR SIR,—I am very truly gratified by the desire of the Committee for superintending the Fox Dinner, that I should take the chair on that occasion. Being obliged to visit Scotland for a different object, I cannot decline so great an honour nor shrink from doing what their wishes convert into a duty.

I hope it will not be accounted presumptuous in me to express my extreme solicitude that the meeting may be respectably attended, and temperately conducted; I feel how much my own defects need to be supplied by the composition of the conduct of the meeting, as I am sure it will be generally felt, that the evil of any failure in either respect would be aggravated by the Presidency of a man who takes an active part in Parliament, and who is now too much detached from his native country to take such a station in the ordinary course of local duty. The hope which you hold out, that Mr. Cranstoun may be persuaded to assist me, is very encouraging. His name alone will make the meeting respectable, as his authority and example cannot fail to make it prudent. If I knew that he had accepted the station of Vice-President my anxiety would be at an end. He would be at once a restraint on indiscretion, and a shield against malevolence. I am on every account sorry that my acquaintance with him is not such as to justify my asking a favour from him. If I had that right, I should now earnestly beg him to lighten what to me is a heavy responsibility. You have probably heard, that I have declined a public dinner at Glasgow; indeed it was my wish that nothing should divert my present Scotch journey of its purely literary character. To the Fox dinner, however, I have sacrificed that wish. I am to be installed at Glasgow on Friday, the 3rd of January. It would be a proof of the delicate regard which I owe to the University to postpone the advertisement of my name as connected with the Fox Dinner to as late a period as is consistent with other objects. As the Committee do me the honour to consult me about the time, I think it necessary to say that my stay in Scotland cannot conveniently be prolonged beyond the 14th or 15th of January. Would Monday, the 13th, be a suitable day for our dinner?—and in that case would Friday, 3rd, or Saturday the 4th, be a sufficient advertisement of my name? If it were thought of any importance the fact might be communicated to our friends in the country by letters some days before.

I shall be very glad to see as soon as convenient a list of the stewards and toasts.

I have not yet determined whether I shall go directly to Glasgow or by way of Edinburgh. This will depend on the wishes of those who do me the honour to accompany me from Edinburgh. If they allow me I shall accept the invitation of Mr. Spiers to live with him at Glasgow. I shall probably bring with me my daughter Fanny,* and my boy, who will then come home from Winchester. If I do go

* Afterwards Mrs. Hensleigh Wedgwood.

to Edinburgh my stay there will only be a day on my way
to Glasgow.

I am, my dear Sir,

Very truly yours,

J. MACKINTOSH.

P.S.—Will you have the kindness to shew this note to
Jeffrey.

[The winter of 1822-1823 was a very severe one and the
illness of their only little boy caused great anxiety to his
parents. The education of their children occupied their
thoughts much, and Mr. Horner took great interest in the
politics of the day, he was a warm advocate of the Whig party,
but was always ready and willing to associate with those who
held different views; nothing could exceed his tolerance, and
willingness to believe that those who differed from him might
do so from the highest motives, and he always maintained
the right of every one to act from their own convictions.

The success of the School of Arts was most gratifying to
him, and about this time, or at an earlier date, the new
school for boys, the *Academy*, was planned, another favourite
child of his. Lord Cockburn, in his Memorials, writes:—
"Leonard Horner and I had often discussed the causes and
the remedies of the decline of classical education in Scotland,
and we were at last satisfied that no adequate improvement
would be effected so long as there was only one great classical
school in Edinburgh, and this one placed under the town
council, and lowered, perhaps necessarily, so as to suit the
wants of a class of boys, to more than two thirds of whom,
classical accomplishment is foreseen to be useless. So, one
day, on the top of one of the Pentlands—emblematic of the
solidity of our foundation, and of the extent of our prospects—
we two resolved to set about the establishment of a new
school. On taking others into council, we found that the

conviction of the inadequacy of the High School was far more general than we supposed. Scott took it up eagerly. The sum of £10,000 was subscribed immediately, and soon afterwards about £2,000 more. We were fiercely opposed, as we expected, by the Town Council. After due discussion and plotting, our contributors finally resolved to proceed, and in 1823 the building was begun. It was opened under the title of the *Edinburgh Academy*, on the 1st of October, 1824, amid a great assemblage of proprietors, pupils, and the public.]

From Mrs. Marcet.

Geneva, *March 23rd*, 1823.

MY DEAR SIR,—I thank you much for your kind remembrance of me, and for all the news you give me of our friends in Scotland, but I am sorry to hear that your little boy is an invalid, and has been so for some time. Heaven grant his health may be restored to him.

I receive no visits till the evening; I am much obliged to such of my best friends who come and spend an hour with me. Frank now goes into society, and he has besides many regular meetings to attend. My girls, too, have a society once a week with other young people to work for the poor, and which would amuse Mary very much; Louisa besides going twice a week to the clergyman who instructs her in religion, which is you know here, an object of much time and attention, previous to receiving the Sacrament, is engaged several hours every day in reading, and making extracts. So that drawing and music necessarily suffer; but it will, I am convinced, strengthen and improve her mind. Frank pursues the study of the law with assiduity, but his favourite pursuit is chemistry; he works with the eldest De la Rive, a young man of distinguished talents and excellent character, so that I am quite happy in their intimacy—did I say happy? alas, my dear friend that is a word I can no longer use, but in the common

routine of language; happiness is no more for me in this world.
I try all I can to occupy my mind, to fill up my time, to excite
some interest for the *head*, for the heart is a dead weight,
dried up and withered; how this world is changed in its
aspect to me; I lived in a garden of Eden, alas! without
knowing it; now it produces nothing but wormwood and
nightshade.   Forgive me, my dear Sir, these unavailing
complaints; it is some consolation to me to unburden my
heart to one of my husband's dearest friends, and on that
consideration I know you will excuse me.

I send this letter by Alexander Prévost, who has taken a
trip to see us before he returns to England; it was a great
consolation to me to see him.   M. De la Rive and Pictet, give
lectures twice a week to the lower classes, in imitation of
your institution, and have a very full and attentive audience,
but this first season their lectures (on chemistry and natural
philosophy) are gratis.   Adieu, my dear Sir, remember me
very kindly to Mrs. Horner and all your family, and to good
kind Mr. Cranstoun with whom we spent a few of our last
happy days.   I know not whether I should be most pleased
or pained to see the Jeffreys here.

Ever affectionately and faithfully,

J. MARCET.

P.S.—Louisa and Sophia beg to be affectionately
remembered to Mary and all your little family.

M. Pictet desires me to say that you have been elected
honorary member of the Society of Arts here, and that your
diploma will be sent you by the first opportunity.

To His Daughter.

London, 6th *October*, 1823.

MY DEAREST MARY.—This will reach you on Thursday,
your birthday, and it will, I fervently trust, find you in perfect
health.   Accept my blessing, and I pray that you may have

a long and happy life, receiving all the good the world has to bestow, for none can merit it more than you do. I know there will be little presents going, and I must not be absent on the occasion, although I cannot be personally present. So I beg that you will take to yourself my copy of Pope's "Poetical Works," in 2 vols., 8vo, bound in pale calf. I think you will find the book, if not in the *Poet's Corner*, in the bookcase between the fireplace and the small room. I shall put your name upon it when I come back, and that I hope will not be very long. I went on Friday to Ham, and spent the day very pleasantly with our friends there.

It was a beautiful day, and I never saw the banks of the Thames in greater beauty, and it was brightened by the autumn tints of the leaves. I walked to Kingston and afterwards to Twickenham, then I paid a visit to Mrs. Drewe and found her and her daughter quite well. Miss Fanny Mackintosh was there, and she and Mrs. Drewe and I walked for nearly two hours in Twickenham meadows. Miss Drewe is going to be married immediately to a Mr. Alderson, a young barrister of great promise.

On Saturday I went to Balham Hill, a little way beyond Clapham, to dine and stay all night with Mr. Alexander Prévost, who returned from Paris last spring, and has got a very pretty house with a good garden, and two fields. His wife was a Mdlle. Martin of Geneva, and has a most agreeable, pretty face, and what is better a very pleasant kind manner, and very good sense. They live very happily with two of his brothers, who are also settled in London, and they have two fine children. I was very sorry to hear a very indifferent account of poor Mrs. Marcet, who suffers terribly from depression of spirits; she has attempted twice to go to Malagny, but in a few days each time she was obliged to leave it. She finds no relief except in moving from place to place, and she even does not take an interest in her children. I hope she will be induced to try the effect of a visit to England.

I sat two hours with the Somervilles last night. She is better than she was, but he, poor man, suffers terribly on the loss of their eldest daughter.

God bless you, my dearest Mary,

Your affectionate father,

LEONARD HORNER.

CHAPTER VIII.

1824—1826.

[The Fox Dinner was held on the 26th January. This year was full of anxiety owing to his little son's long illness, which proved fatal in the autumn—but he occupied himself, besides his interest in politics, with the School of Arts, and with the organization of the new classical school, "the Edinburgh Academy." He wrote to Dr. Parr for advice concerning this school and received from him the following reply.]

From Dr. Parr.

Hatton, *February* 20*th,* 1824.

DEAR AND MOST RESPECTED Mr. HORNER,—I thank you for your friendly and interesting letter, and I have read attentively the pamphlet which accompanied it; the schoolmasters, and ecclesiastics of South Britain, will smile at the minuteness of your details, and the accuracy of your calculations; but they are well adapted to the subject, and worthy of a considerate people, who know how to avoid the misapplications of abundance, and the inconvenience of deficiency. Upon the choice of books for your upper forms, I cannot now offer you any advice, but I will hereafter tell you my opinion freely, when your plan is more mature, and your Rector is appointed. For reasons which I need not specify, you would do well 'to get your Rector from England. But I tell you *in limine,* that the allowance you mention is too scanty, if you choose to employ a thorough scholar. You must provide for him a well furnished house, and his annual salary must be £500, and these advantages, with the cheapness of living at Edinburgh, may be allurements sufficient to make an English scholar exchange his own country for yours. I suspect that your project will not create much interest among my countrymen. You may give publicity to your Academy, by sending your pamphlet to our public schools, and our two universities; it differs so widely from what is pursued in our seminaries of education, that I doubt whether you can obtain any useful suggestions. This point you must settle for yourselves; however, I will send you a list of the persons whom you may consult.*

Dear Mr. Horner, I have sent you a very full list, and if you think it worth your while to apply to any of these persons, you will do well to employ your best sense, and your best style in a circular letter. I forebode little good from the application. Pray remember me kindly and respectfully to your lady, to your father and mother and to all your family, to Mr. Alison, to Mr. Pillans, to Mr. Dunbar, to Dr. Jameson, to Mr. Jeffrey, to his brother-in-law, Mr. Morehead, to Mr. Playfair, to Dr. Thomson, and his lady, to the widow of Dr. Gregory, to Mr. Murray, to Mr. Thomson, keeper of the Archives, and to Mr. H. Mackenzie, whose life of John Home delights me. I was charmed with a biographical work from Mr. Thomson.† You see, dear Sir, how well I remember, how highly I respect, the literary worthies of Edinburgh.—I am, with great esteem and regard, dear Leonard,

> Your friend and obedient, and humble servant,
>
> SAMUEL PARR.

[On the 1st October, 1824, the school, on the establishment of which Mr. Horner had devoted so much time and attention, was opened under the title of the Edinburgh Academy, amid a great assemblage of proprietors, pupils, and the public.

---

* Here follows the list.

† The life of Lady Grisell Baillie.

Henry Cockburn writes thus :—" We had a good prayer by
Sir Harry Moncrieff, and speeches by Scott and old Henry
Mackenzie ; and an important day for education in Scotland,
in reference to the middle and upper classes. Mackenzie's
vigour was delightful, though above eighty, he made an
animated address, exulting in the rise of a new school upon a
reformed system."

Mr. Horner was one of the original directors of the Academy,
and took the most active part in the management of it, during
the three following years in which he remained in Edinburgh.
Amidst his active exertions for the service of his native town,
and his various domestic anxieties, he always found his
favourite study, geology, a constant resource in his leisure
hours.

He derived also much pleasure from a wide circle of
friends and acquaintances in Edinburgh, most of them Whig
lawyers. William and John Murray, Henry Cockburn, Francis
Jeffrey, Thomas Thomson, Andrew Rutherford, Thomas
Maitland, Sir James Gibson Craig, and his two sons, William
and James, and many others. There were a great
many Italian refugees in Edinburgh this season, some
of them very agreeable, most especially two brothers,
Baron Ugoni and Count Ugoni, they always received a
welcome at Mr. Horner's house, but there were some with
whom his family became very intimate, M. Viaris, a
Piedmontese, had been a distinguished soldier in Napoleon's
army, had been through many of his campaigns, including
the Russian one, but on Napoleon's fall, had taken refuge
with his family at Geneva, which he was obliged to quit by
the desire of the French Government.

At Geneva, he and his wife had made the acquaintance of
Lady Anna Maria Elliot, who recommended them to go to
Edinburgh, and sent letters of introduction to them there.
M. Viaris was a very handsome man, with gentle, agreeable

manners. Mdme. Viaris was a German, a very clever, superior
woman. They had a son of eleven and three younger daughters.
Their eldest daughter afterwards married a distinguished
Russian, Mr. Nicholas Tourgueneff, cousin of the novelist of
that name.

On the 29th July the first public exhibition day of the
Edinburgh Academy took place. The third prize of the
Rector's class was given to Archibald Tait, since Master of
Rugby School, Dean of Carlisle, and Archbishop of Canterbury.
He had also a prize for the best English essay, which was on
the advantages derived from the study of history. He was
then only 13 years old, as he was born on the 31st December,
1811.

Mr. and Mrs. Horner paid a visit in August to Mrs. Horner's
cousin, Sir Robert Bateson, in Ireland.]

To his Daughter.

Belvoir Park, Belfast, *August 28th*, 1825.

My DEAREST MARY,—We are just returned from Newton
Breda Church. We hope to be with you next Sunday.

Yesterday your Mamma with Mrs. Bateson and Sir Robert,
went over to Moira by Lisburn; Sir Robert is Lord of the
Manor there, and had so much to do that they remained till
nearly seven o'clock. Your Mamma came home quite delighted
with the beauty of the country. I did not go, because I was
engaged to dine with a Mr. Lyons, where I met a very large
party of men, Lord Donegal, Sir Stephen May, and sundry
others. Doctor McDonnell lent me his gig, and I drove up
to Belvoir last night, which was a very fine one, as the moon was
shining bright. We have had an invitation to go to Cultra
on Tuesday, to breakfast there, go on board a Swedish 74
lying in the Loch, dine and stay all night, but Lady Bateson
cannot leave her baby, and as I have a great dislike to water

excursions we do not go. We are to spend one day with Dr. McDonnell, as he is very anxious to get acquainted with your Mamma, and we shall sleep at his house on Wednesday evening. We have, hitherto, had most charming weather, but it looks to-day. as if it were going to break, and I fear we shall not have the luck of good weather for our voyage.

Since I began to write to you I have received your very agreeable and excellent letter of last Friday, and it makes us very happy to find that all the treasures we left behind us, are so well. We are very glad to hear that your uncle has at last appeared, and we hope that he will stay with us as long as possible.

How very kind Mr.* and Mrs. Cockburn have been to you, I daresay you passed a very merry day there; but had I been there, I should have had great fears about your remaining so long upon the bowling green at this time of the year, and I hope that neither you nor your cousins feel any bad effects from it. As to your not studying so much as you intended, I am not at all surprised, nor do I think there is any harm done. In order to keep your mind at ease about your progress in your studies, lay out a plan for some months before you, and in forming that plan, take largely into account those interruptions from society and other causes, which you will inevitably meet with. By reducing this plan to writing and occasionally looking at it, it will be a great help to you. Do not allow yourself to look back vaguely on the past, and accuse yourself of mis-spent time; when you are undressing for bed, review each day, and if you are satisfied upon the close review, that what you did amiss arose from circumstances which you could not avoid, you need have no cause for self-reproach.

Monday, 29th August.—Dr. McDonnell dined here yesterday, and I am glad that your Mamma likes him, for it is essential to my happiness that she should like all those I esteem.

* Afterwards Lord Cockburn.

He was very agreeable, and without being formal, or pedantic,
he is one of those persons who leave an impression that the
time passed in their society has been rationally and agreeably
occupied. . . . . . If the day is fine to-morrow, I
mean to take your Mamma over to Lisburn to see the famous
damask manufactory of Mr. Coulson, and the bleaching
grounds of Mr. Williamson at Lambeg. I had at one time
intended to have taken your Mamma to the Giant's Causeway,
but we think it better to put it off, because my business, as
well as my inclination, will very probably bring me to this
country again, either next year or the year following, when
we shall, I trust, have you with us. God bless you, my
dearest Mary.

Your very affectionate father,
LEONARD HORNER.

[In 1825 Mr. Horner was elected a member of the Friday
Club.

This Club arose in Edinburgh, in 1803, and was so called
from the day on which it first used to meet. The idea,
according to Lord Jeffrey, was Walter Scott's. It was
entirely of a literary and social character, without any limitation
of numbers ; it was open to any person generally residing in
Edinburgh, who was supposed to combine a taste for learning
and science.

It was late in this year that the Whigs of Edinburgh gave
a dinner to Mr. Joseph Hume, and requested Mr. L. Horner
to take the chair on that occasion.]

From Francis Jeffrey.]

Craigcrook, Edinburgh, November, 1825.

MY DEAR HORNER,—You would not have found Thomas the
rhymer here to-day,* and though he dines here to-morrow for

* Thomas Moore, the poet.

the last time, I am afraid the smallness of my table must
prevent me from asking you and your ladies to place yourselves
at it.   If you are not frightened at the length of the journey
in the dark, Mrs. Jeffrey and I would be most happy to see
you here early in the evening—say soon after eight—when
you might hear him sing, and if you do venture on this, I
hope you will bring all, or some of your beauties, with
you.

I am afraid there is not the least chance of his staying till
the 18th, as he seems resolved to set off on Wednesday next,
or Thursday at the latest.   Neither *entre nous* do I think that
he would stay a day to do honour to Joseph,* even if he came
at the head of Pharoah's household as of old. I will come to you
with pleasure, and I shall be happy to lend you any feeble aid
I can, as to toasts, &c., though that is not the department in
which I profess to shine.

I think it was quite right to place you in the chair, and
very right and meritorious in you to go into it, and I hope
you believe that with these sentiments, I should despise
myself, if I did not do all that is in my power to support you
in it, and to lighten your labours with a view to it.

Believe me always, very faithfully yours,

F. JEFFREY.

Letter From Lord Cockburn To Mrs. Horner.

19*th November.*

MY DEAR MRS. HORNER,—I cannot deny myself the pleasure
of assuring you, in strictest truth, and not the least in
compliment, that Horner acquitted himself last night to the
admiration of everybody.   He is the very best chairman at
such a meeting I have ever yet seen.   He certainly is a
capital chairman, particularly in the two most difficult parts

* Joseph Hume, Esq.

of that curious art, the first discourse, and the general
management. I was very nearly getting up to propose that
he should be appointed chairman general for Edinburgh.
But I was afraid that our rivals of the Corporation might be
uneasy. I again assure you that his best friend could never
wish him to do better than he did last night.

Ever yours,
H. COCKBURN.

To his Daughter.

London, March 21st, 1826.

MY DEAREST MARY,—I met last week, by mere accident,
Mr. Hedworth Lambton. I breakfasted with him on Sunday
morning. He has seen a great deal of Italy and Sicily, and has
made some very interesting collections of works of art. He
has so many, that he has hired a room to 'lay them out, and
I am going to see them. He says that he must spend next
winter in Edinburgh and thinks it the most delightful place
he knows. He has very vivid and very proper recollections
of the Baron of Bonaly* and his domain.

I went to Hampstead with Mr. Whishaw to visit the Mallets,
and passed a most agreeable day, as well there, as in the
carriage with Whishaw. We had only Warburton and John
Prevost, and much good conversation. Lucy Aikin came to
tea. Mr. and Mrs. Mallet are both well. Warburton you
know is a great favourite of mine, he is a very capital man.

Yesterday I dined at the Duke of Somerset's, we had Prince
Leopold, Lord Prudhoe (brother of the Duke of Northumber-
land), Lord Hervey (Lord Bristol's eldest son), Lord Morpeth,
Sir Stamford Raffles, Captain Parry, and Mr. Sharp. Lord
Seymour, the Duke's eldest son, was at home from Christ
Church. It was a very pleasant party, and was not rendered
in the least degree stiff by the presence of His Royal Highness,

* Lord Cockburn.

who seems a good natured, unaffected, and fairly sensible
man, who did not talk more than his share. I was very
favourably impressed with the conversation of the young
nobles, particularly with that of Lord Prudhoe—the tone was
such as to show that they were accomplished, and desired to
be thought well of by so excellent a critic as Sharp. His conversation is always instructive and agreeable. Sir Stamford
Raffles has been been a great deal, you know, in the East
Indian Archipelago, and is a very intelligent person. You
have probably heard of his misfortune of losing all his MSS.
and collections, by the burning of the ship on which he had
embarked on his return from his Government, 24 hours after
they had sailed. They scarcely escaped with their lives. I
had some conversation with Captain Parry; he has no
intention of going upon the same expedition again, but he has
a great desire to put his foot actually on the end of the earth's
axis, at the North Pole, over the ice from Spitzbergen, in
boats made of skins. The great difficulty is, to transport
provisions.

I have seen Mr. Barker several times, and found him
occupied in many literary schemes; I thought it would do
him a great service to introduce him to that worthy Mecænas
of all men of true merit, Whishaw—and we breakfasted with
him this morning, and passed a very agreeable hour. He
shewed us some interesting literary curiosities, particularly a
collection of Rousseau's letters in his own hand-writing, to the
Marquis Mirabeau, the father of the person who distinguished
himself so much in the French Revolution. Whishaw will be
very kind to Mr. Barker, and will introduce him to Sir James
Mackintosh. He has an historical work in contemplation, an
edition of Hume, with notes to serve as an exposition of the
errors into which that seductive historian is too apt to lead
his readers, errors which have been very ably but drily pointed
out by Brodie.

God bless my dearest Mary, my affectionate love to your Mamma and sisters.

Your affectionate father,

LEONARD HORNER.

[In 1826 it was determined to found a College in London, to be called the University of London. The Council first formed to establish it, included many remarkable names, among them:—The Hon. James Abercromby, Lord Auckland, Lord Dudley and Ward, Alexander Baring, Thomas Campbell, Henry Brougham, Joseph Hume, Lord Lansdowne, Zachary Macaulay, Sir James Mackintosh, James Mill, Duke of Norfolk, Lord John Russell, William Locke, Henry Warburton, John Whishaw, Isaac Lyon Goldsmid, &c., &c.

The Prospectus which was headed by these names stated, "That the City of London is nearly equal in population, and far superior in wealth, to each of the kingdoms of Denmark, Saxony, Hanover, and Wurtemburg; every one of which, has at least one flourishing University. It may be safely affirmed, that there is no equal number of youths in any other place, of whom so large a portion feel the want of liberal education, are so well qualified for it, could so easily obtain all its advantages at home, and are so little able to go in quest of them elsewhere. The exclusion of so great a body of intelligent youth, intended for the most important occupations in society, from the higher means of liberal education, is a defect in our institutions, which, if it were not become familiar by its long prevalence, would offend every reasonable mind.

"In a word, London, which for intelligence and wealth as well as numbers, may fairly be deemed the first city in the civilized world, is at once the place which most needs a University, and the only great capital which has none."*

The Council, through Mr. Henry Brougham, requested Mr. Horner to undertake the office of Head or Warden of the new college, which after due consideration, he finally agreed to.]

* For further details see the printed prospectus of that time.

CHAPTER IX.

1827—1829.

To his Daughter.

London, *15th April,* 1827.

MY DEAREST FRANCES,—I assure you my thoughts are very
often at home, and I long very much to be there. Knowing,
however, that I must stay, I endeavour to make the time
pass as agreeably as I can, and that is not very difficult, for
London is a very delightful place to me, where I have a great
many friends. The only sight I have ever seen is the new
panorama of Geneva, I was delighted with it, and my desire
to go there is greater than ever. The view is taken from that
part of the lake close to the town, which is opposite the
Douane, and on the left the Salève, and Mont Blanc in the
distance. I have been to see a most beautiful collection
of shells. They belong to Mr. Broderip, who is one of the
Secretaries of the Geological Society, and is very learned in
that branch of Natural History, and has one of the finest
collections in England. I never saw any shells at all to com-
pare to them in point of beauty, and it was not their forms and
colours alone that pleased me, for he pointed out many of
those parts of the structure known only to those who have
attentively considered the subject, which exhibit those
beautiful contrivances with which every department of nature
abounds, when we look narrowly into her works. I was quite
surprised at the value of some of the shells. He shewed me
one for which he had given forty guineas, many that he had
given twenty, fifteen, ten and such prices. His collection is
worth several thousand pounds. His knowledge is not
confined to shells, he is an excellent geologist, has many

beautiful books, and is a very agreeable and entertaining person. He is one of the Police Magistrates. He shewed me a curious copy of Walton's "Angler," with a print of every person mentioned in the book, as well as of almost every place. I hope to put our shells in systematic order, for I mean to make myself acquainted with the names of shells, for my ignorance of them is a great impediment to me in many of my geological inquiries.

Your affectionate father,
LEONARD HORNER.

To his Daughter.

London, 21st April, 1827.

MY DEAREST SUSAN,—I went yesterday to see the Tunnel under the Thames. The people of Kent who have business in the eastern parts of London on the left bank of the Thames, find it very inconvenient to go so much out of their way as they must do by going over London Bridge, and as it is impossible to make a bridge so low down the river, on account of the shipping, a very ingenious engineer, Mr. Brunel, said, "why not dig a hole <i>under</i> the Thames, I assure you it is very practicable." His reasons were so good, that he found a great many people willing to come forward with the necessary money, they expecting to be paid by a toll to be paid by passengers. The spot they chose was very near Rotherhithe Church, a place you may have heard of, as that renowned traveller, Mr. Lemuel Gulliver, was born there. They began on the south side by digging a great pit or shaft, fifty feet in diameter, which is lined with bricks. After getting down to such a depth as would leave a strong roof between them and the river, they began to make the tunnel in a line nearly horizontal, and when they have gone quite under the water, another shaft will be dug on the north side of the river. There is a very easy descent by an excellent staircase

down the shaft, and a more striking sight I never witnessed
than when I got to the bottom and saw the long vista of the
tunnel lighted up with gas. They complete the arch as they
go on, and for four hundred and fifty feet it was as dry as the
room you are sitting in. Think of my being under the
Thames with great ships sailing over my head. Having a
letter from the engineer, we were conducted by special favour
to the spot where the miners were at work, and a most
curious and interesting sight it was; the process however, is
too complicated for me to explain to you, but there is no
doubt entertained of its being completed, and when it is so,
it will be one of the greatest efforts of human ingenuity that
exists in the world. The length of the Tunnel from shaft to
shaft, will be 1,300 feet; I went with Mr. Decimus Burton,
and Captain Pringle of the Royal Engineers. We called for
the latter gentleman at the Tower, where we saw the operations
going on there, about making a great map of England,
Scotland and Ireland.

I expect to breakfast to-morrow morning with Mr.
Humboldt, the great traveller, who went over the Andes.
Give my love to all your sisters, and to Miss Parker give my
best regards.

I am, my dear Susan,

Your very affectionate father,

LEONARD HORNER.

To his Daughter.

London, 1*st May*, 1827.

MY DEAREST MARY,—I went yesterday to the laying of the
Foundation Stone of the London University. I refer you to
the newspapers, sent by this post, for some of the details, but
they have failed to give Dr. Lushington's speech, and I cannot
supply the defect. It was a most beautiful day, and there was

a great crowd of people, and everything went off as well as
could be wished. I am sorry that the newspapers give so
meagre an account of the speeches at the dinner, for both
Brougham's and Lord Lansdowne's speeches were particularly
good. There was an immense assembly, nearly five hundred
people were in the room, and more than two hundred
applicants for tickets, could not be supplied. Dr. Fitton and
I went together, and by Mr. Loch's kindness we got very good
places. The only thing uncomfortable was the terrible heat.
The whole proceedings passed off admirably, and will, I have
no doubt, produce a very favourable impression. I never saw
a meeting where there was so large a proportion of people of
station and property. I met Winthrop Praed there, who
came up from Eton on purpose to attend the meeting, with
his friend Orde, the son of the member for the county of
Durham. He is looking in very good health, much more so
than when I saw him last. I am sorry to find that his family
will not be in town for a fortnight, for I hope to be gone before
that time. My letters lately have been so full of the interesting
political events that have happened for the last fortnight,
that I believe I have not said much of where I have been
spending my time. I dined at Hallam's last Thursday, and
had a very pleasant party, Whishaw and Loch. In the
evening Whishaw and I went to Lady Minto's. On Friday I
was at dinner at Mr. Irvine's, one of the masters of the
Charter House, we had Sir Pulteney and Lady Malcolm, he
is a very hearty, and agreeable old gentleman, Dr. Buckland
and his wife, and the new Bishop of Calcutta, and his wife—
Mr. James is his name, and I knew him before. You will
find an interesting account of his predecessor, Dr. Reginald
Heber, in the last number of the *Quarterly*, which you can
borrow from Cockburn, as I think he takes it. I dined *en
famille* with the Bells on Saturday, and found him in pretty
good spirits. . . . . . We had an exceedingly pleasant

party at William Murray's in the Temple on Sunday;
Count Flahault, Kennedy, Mr. Phillips, Sir James Graham of
Netherby, William Brougham, and Rutherford. We kept it
up to a very late hour, and at half-past eleven we were
walking in the Temple Gardens, the air very mild, and the
sky quite brilliant. Count Flahault and Lady Keith are
going to Paris. Rutherford had just come from thence, and
is at this moment on his road to Edinburgh. He said he
should call and let your mamma know that he has seen me.
To-day I am going to dine with Dr. Fitton, to meet Baron
Humboldt. On Thursday I expect to be at Hampstead,
dining with the Mallets, to meet the Prévosts. On Friday is
the Geological Society Club, and meeting in the evening, and
I have just received an invitation to dine with the Marshalls
on Sunday. . . . . . . I am very glad to find you
continue to like drawing so much, as I am sure it will be a
great source of interest to you, and now that the summer is
approaching, I hope you will be able to exercise yourself in
the great object, that of drawing from nature. I wish you to
go on for another quarter at the Drawing Institution, for
even if the event we have in contemplation, should take
place,* which is certainly far from a settled point, although I
think probable, I daresay you will have ample time to finish
one quarter. It will confirm the good you have got in that
previous half year. I shall depend much upon your assistance
in my project of getting acquainted with the forms of shells,
and what we have already, will be a very good exercise. It
has become a very important element in geological knowledge,
for shells seem to determine many things in the history of
the rocks in which they are found. They are the coins and
medals of the world before the flood. Not that I would have
you understand by that phrase, that the flood had anything to do with locking them up in those cabinets in which

* That of moving to London.

they have been preserved to excite our wonder. I hope Jeffrey has got home again quite well, remember me to them in the kindest manner. Say everything that is kind to the excellent family of Viaris, from me.

I mean to bring you an edition of Dante, that has been lately published here, by an accomplished Italian of the name of Rosetti, with many valuable notes, and the verse turned into modern and very elegant Italian, in order to assist one in understanding that very obscure and difficult author. I believe that the "Inferno" is only yet published, but that the author means to go on with the rest of the Divina Comedia.

Remember me most particularly to Cockburn, and to all the other good people.

I am, my dearest Mary,

Your very affectionate father,

LEONARD HORNER.

To the Right Hon. the Lord Auckland.

41 Russell Street, *May 14th*, 1827.

MY DEAR LORD,—I received, on Saturday evening, a letter from Mr. Coates, enclosing a copy of the resolutions of the Council of the University of London, in which they have done me the honour of proposing that I should accept the important office of Secretary. I am very sensible of the great distinction conferred upon me by such an offer, and the more especially when I consider the very eminent persons who compose that Council. I am quite willing to undertake all the duties mentioned in the resolution, as well as that general superintendence which has been pointed out to me in conversations I have held with members of the Council. It is proper, however, for me to state, that while I do not object to the name of Secretary, if the Council think that is the

most appropriate term, my understanding of the nature of
the office is this—that as the organ of the Council, I am to
possess that authority over the professors, and the various
officers of the establishment, which is usually vested in the
Principal or acting Head of other Academical Institutions—
that I am not to hold a subordinate situation, or be under
the control or direction of any other power than the Council.
The proprietors may see reason to institute some high
dignified office, such as Chancellor; but the person holding
that office of honour, would not, of course, interfere with the
management.

I accept the remuneration proposed to be annexed to the
office, when the academical duties commence. . . . . .
I beg that the Council will take into consideration the
observations I have taken the liberty to state, relative to my
understanding of the nature of the office, the adoption of the
descending scale, and the remuneration for the intermediate
period; and I request that they will favour me with their
answer at their earliest convenience.

> I have the honour to be, my dear Lord,
> Most faithfully yours,
> LEONARD HORNER.

The Right Hon. Lord Auckland.

From Mrs. Marcet to Mrs. Horner.

Aix-in-Savoy, *June 16th*, 1827.

I HAVE just received, my dear Mrs. Horner, the news of Mr.
Horner's appointment with that delight which it must give
to all his friends who wish to see him once more settled
among them in London, who feel a sincere gratification in
seeing merit rewarded, and one so highly and generally
valued, placed in so public and distinguished a situation.
Independent of my sincere and warm friendship for Mr.

Horner, I view it in the light of promoting the cause of
virtue and improvement, that one who has with so much
disinterested zeal, devoted his time and his talents to further
the general diffusion of knowledge, should obtain an eminence
so peculiarly appropriate. I hope, my dear Mrs. Horner,
personally to congratulate you next winter, and among your
numerous London friends to rejoice in your restoration to the
great metropolis. I thank you for your kind letter on the
marriage of my son; if I had not previously considered him as
determined on settling at Geneva, I might have had some
regret at his rivetting the chain which binds him to his
father's country, but that point being in a measure settled,
the individual he has chosen is everything I can wish. Frank,
who is now in England, introducing his darling bride to her
new relatives, wrote me word of Mr. Horner's election, and
said he was coming with his family to London; but as you
are coming for *good and all*, you must doubtless have much to
arrange previous to your departure, and I fear will not
arrive in England before F. Marcet's departure; we are
impatient to have them back, that we may spend some
months together, before I set off for England with my
daughters. We are at present at the baths of Aix-in-Savoy,
for the benefit of my brother William's* health. We are
making such considerable repairs and improvements at
Malagny, that we are unable to inhabit the house this summer,
fortunately we have a smaller one on the premises, sufficient
for our provisional accommodation, which will enable us to
overlook the workmen.

Ever affectionately yours,

T. MARCET.

[Early in October Mr. Horner removed his family to
London when he became Warden of the London University.]

* Mr. Haldimand.

From Mr. Henry Cockburn* to Mrs. Horner.

Bonaly, 11*th October*, 1827.

My Dear Mrs. Horner,—A letter from Walker Street told
us of your prosperous voyage, and safe arrival in London.
But we long to hear from you yourself.   Remember that it is
now chiefly by letters that our acquaintance can be maintained,
and you can scarcely suppose how time, distance and
negligence make the best acquaintances fall asunder.   Is all
the furniture arranged yet?   This will keep you all busy for
some time, and by so doing, will help to keep you from many
—at least some—uneasy thoughts.   Your chairs and tables
if they have been with you long, must be travelled articles, I
suppose they were shone upon by the pure sun of Whitehouse;
covered by the house in King Street; descended the
adventurous but comfortable Lane; and now are surprised to
find themselves in the capital of England.   When shall I see
one of the old parties sitting upon, and round them?   I have
often thought that when people were well set, it is a pity
that Providence won't glue them where they are.   But then
think how we would look if we were to discover some eternal
friend beside us whom we did not like.   Pray do mention us
sometimes to the younger children.   It will require every
thing to keep us at all in their memories.   Mary I know, and
I trust Frances and Susan, will never forget us.

It has been a sad breaking up of many of my visions for
hereafter, but there can be no doubt of its wisdom, and, if we
choose, the change may be made a source of much new
happiness, and mutual friendship.   I sympathise with Leonard's
more congenial employment, and envy his means of increased
reputation and usefulness.   I trust he will never renounce
his care of the public cause of Edinburgh—to which he has
already contributed so much.   He must know that all the

* Afterwards Lord Cockburn.

good we have ever done, has been through our friends in
London, and the presence of one who knows our circumstances
so well, and has such access as he has to people of influence
there, may, without in the least taking him out of his
proper line, be of immense benefit. I find that our boys have
taken a very delicate and effectual method of consoling the
house in Walker Street* for your absence, they have established
themselves in it by inviting themselves to tea. I know
nothing more interesting than the old gentleman's reminiscences
of olden times, and it is impossible not to admire what he has
done for his children, particularly in forwarding Frank, and
now—in his old age—so magnanimously parting with
Leonard. The children here often ask about you, and can't
understand why you have gone away. They are all well,
though, from Elizabeth upwards, much annoyed by the rain
which confines them within this large and roomy tenement.†
This day the burn is high and red, the sure signs of sudden
rain, near a hill. I don't know that I ever saw heavier or
more steady torrents, I have been going over the terrace,
transplanting evergreens, and I daresay it is all for my sake.
I have resolved *not* to take the fields to the east of me,
they are too dear ; and if my own shrubs thrive as I intend,
I shall soon be protected by their foliage from all the harm
that a neighbour can do me. I mean to be in perfect beauty
in the year 1835, by which time all the trees will be ten years
old, and the new house will be finished.

Keep Leonard and yourselves as much in the habit of
walking as a London life admits of ; else where are the merry
rural breakfasts at Habbies How, or the glorious lung-
opening, foot-trying prospects from Cape Law and Carnethy ?
I need not advise you to preserve in your souls a due relish
for these, and such scenes and occupations, for I cannot

* At Mr. Horner, Sen.

† Bonaly at that time was a very small house.

suppose you to lose it. But even though you should, nay,
though you should even go the horrid length of allowing any
portion of your love for me to decay from separation, I don't mean
to repay you in kind. I have often lamented of late years that
circumstances—chiefly my abhorrence of strangers—kept us,
through within a few yards of each other, so long unacquainted,
and I seriously am resolved to make up for the loss, by
increased tenacity now that I have got you. Your absence
makes, and will make, a sad blank this autumn and next
winter, in so much, that since Richardson went away, twenty-
one years ago, no such gap has been made in my habits and
society. I don't mention this for the silliness of despondency,
for since, for prospective purposes, it was right, I rejoice
that the sacrifice was made, but to let you know that you
may ever depend upon me, and that I know we may with
equal constancy depend ever upon you and yours. God bless
you. Write soon.

Ever,
H. Cockburn.

1828.

[In the spring of this year Mr. Horner, sen., with his
daughter, came to London to visit his son and his family, and
during the summer, a house was taken in Leatherhead, Surrey,
where the family adjourned for rest and recreation, Mr. L.
Horner joining them as often as his business permitted; for
already the responsibility and harassing cares of his office
began to tell on his health. The summer passed very happily
in that beautiful neighbourhood, and they had visits from Mr.
and Mrs. Mallet with their three little sons, and Mr. Henry
Cockburn arrived from Scotland; with him they went to
Wotton, Evelyn's home, and the little church associated
pleasantly to Mr. Horner's mind that his great-uncle, the
Rev. Thomas Broughton, had been the revered pastor of that

parish. While at Leatherhead, Mr. and Mrs. Horner and
their two eldest daughters went to Walton-on-Thames to visit
Mr. James Mill, the historian of India, where they met his
son, known later as the eminent John Stuart Mill.]

From Lord Cockburn.

14, Charlotte Square, Edinburgh, *6th July*, 1828.

MY DEAR MRS. HORNER,—I have written to Richardson
to-day, proposing a project—the chief object of which is to do
good to your worthy spouse's health, which I am satisfied has
been impaired solely by over-working, want of exercise, and
of *me*. The scheme is, that he and Richardson shall pass
August with me, in going through Wales, or the West of
England, or anywhere else they like. I shall join them
wherever they please, and shall be all submission to whatever they wish (provided I always get them to wish what I
think right). At one end of the journey or other, I shall hope
to be allowed to see you and Mary, who I suppose it is in
vain to invite to join the party? What do you say to this?
If it be good for Leonard, what can it be bad for? If it be
bad for him, it is good for nothing.

Ever yours,

H. COCKBURN.

From the same.

Hampstead, *August*, 1828.

MY DEAR MRS. HORNER,—I won't, and can't, say anything
about our three last pleasant days; but I hope that many
such days are reserved to us. I am sure that if we always
prove true to each other, they must be so, even " amidst the
changes and chances of this world."

The particular purpose of my writing to you now, is to
induce you to induce your spouse to go with Richardson and

me, on Tuesday the 19th, to the Isle of Wight. Such an
expedition would do his spirits much good, which, besides the
pleasure of his company, is the main reason that makes me
anxious about it. To remove all pretence of wasting time, I
have proposed to him that he should so far alter his plan, as
to go to Leatherhead on Saturday instead of Friday—to go
to Mr. Mills on Sunday, and to London on Monday. This
will let him start with us on Tuesday at twelve, free of
remorse. He says he is *anxious* to go, and that he *thinks* he
will go, and that he will if he *can;* which you know means,
that he neither wishes to go, nor can go, nor will go. In
order to whet his blunted purpose, I write to you. The idle-
ness and play of our Tuesday, Wednesday, Thursday and
Friday, will create a soul under his ribs; so, as he, like other
good men, is entirely in the hands of his wife, do you immediately write a letter to him, saying " My Dearest Leonard,
Mr. Cockburn is always right; go—we implore you—go with
him and Mr. Richardson, on the outside of the Portsmouth
coach, to the Isle of Wight, and return to us with renewed
heart and vigour. Your affectionate wife, Anne Horner."
Now do write this—and he can't resist.

I shall have a full communication with you and Mary soon.
But the above admits of no delay. In the meantime, God
bless you both and all.

Ever yours faithfully,

H. Cockburn.

[Mr. Horner was unable to go. His health rapidly declined..
Sir James Clark, his kind friend and physician, always gave
him much comfort, but early in the winter, he was advised
to spend a few weeks at Brighton, and he also went with his
wife to visit his friends Mr. and Mrs. Tertius Galton, near
Birmingham.]

To His Daughter.

Birmingham, *New Year's Day*, 1829.

While I am waiting in the Bank for Mr. Galton to go home,
I cannot employ my time so happily as by writing to my
dearest Mary, to send my own and her dear mother's most
fervent wishes for many happy returns of this day to herself,
and our dearest Frances. This is the day you were to go to
Ham,* and I hope that you have arrived there in safety,
and that Mr. Nicholson was duly sensible how great were the
treasures which he had charge of.

Yesterday your Mamma got both your letters, the contents
of which gave us great comfort, not only by learning that
you and your sisters were well, but the account you gave us
of your judicious arrangements in our absence.

I am beginning to experience some benefit from the country
air, and from the quiet life I am leading. I have suffered
less from depression of spirits, and feel somewhat stronger,
but I must not boast too soon. I trust by a continuance of
this life for another fortnight, I shall feel a more decided
improvement, your dear Mamma is, I am happy to say,
quite well, and has had no headache for a week. The
Galtons have been desirous of giving us amusement by
inviting people to meet us, but we have preferred living with
them alone on several accounts.

The only place I have been to was this morning, Mr. Galton
and I left home soon after eight, to breakfast with Mr. James
Watt, who lives at Aston Hall, about three miles from the
Larches. Aston Hall is a very fine baronial mansion, built
by the family of Holt, in the reign of James I. It is in
the form of the letter E, and resembles Holland House a good
deal. There are regular records of the Lords of the Manor
since the Conquest. It still belongs to the family of Holt,

* To George T. Nicholson's, Esq.

Q

but Mr. Watt has obtained a very long lease of it; and has
put it in repair, and furnished it with great taste.   Charles I.
took refuge in it, and it was besieged by Oliver Cromwell.
We saw very many places on the staircase, where the shots
had penetrated, and one cannon ball which struck it is
preserved.   There is a very fine gallery, one hundred and
thirty-six feet long.   Mr. Watt has, besides plaçing in it
appropriate furniture, made a most excellent library, and on
casting my eye over the books, I have seldom seen a
better or more enviable selection.   He has also some good
pictures, and good copies of celebrated pictures.

We found Mr. Watt just going to breakfast, with the old
Admiral, Sir Isaac Coffin, a humorous old gentleman, and
we passed two hours very agreeably.

Pray, my dearest Mary, do you and Frances take good care
of yourselves; remember how much our happiness will be
increased during our separation from you, if we feel assured
that you and she are careful of your health.   Give our
kindest regards to Mr. and Mrs. Nicholson.

Believe me ever, dearest Mary,

Your most affectionate father,

LEONARD HORNER.

———

From Francis Jeffrey.

Edinburgh, *7th February,* 1829.

MY DEAR HORNER,—It is always a pleasure to hear from
you, but I should have been better pleased if you had said
that your health was quite restored.   I know you are a good
patient, and am quite confident that a little longer
continuation of care will set you fairly up again.   Your
father at one time made us all very uneasy about you.   But
he seems now to be satisfied that there is no cause of
alarm.

I rejoice in your Academic prosperity, If you get creditably through this novitiate, I have no doubt that you will increase and multiply. I took a bet that you would have five hundred before the end of January, and I am inclined to take another that you will have seven hundred and fifty next January, and one thousand the year after. I wish you could get rid of the Academic funds altogether, and let all the subscriptions be turned into free donations. I should think that four-fifths of subscribers would do this at once, for what can they care about four pounds a year? and it would relieve you of a dead weight.

I hope we are to see you in the course of the summer. I have some notion that we may come up to town in April, but we shall see. Tell Mary that she must not-forget me, and do not let Mrs. Horner or the younger ones forget that they are half Scotch, however humiliating it may appear.

We are all quite well except that I have still some infirmity in the trachea. Fullarton's appointment is the best possible, except that Bell* has been unhandsomely used, and I am afraid he feels it very keenly.

We are all on the tiptoe for the speech† which is expected this evening, but I am going to Oxenford to dinner, and shall not see it till night. Things look better, I really think, than ever, though it is not my disposition to be very sanguine. Let me hear from you now and then when you are at leisure.

Ever very faithfully yours,

F. JEFFREY.

[Mr. Horner, with his family, went to his father's house in Edinburgh in June of this year. He attended the General Meeting of the Proprietors of the Edinburgh Academy, and

George Joseph Bell, an eminent lawyer.

† The King's speech was read on the 5th February, but did not reach Edinburgh, in those days, till the 7th. It was in this speech, that, through the influence of Mr. Peel, who was Home Secretary, came the recommendation of the King to Parliament, to consider whether the civil disabilities of the Catholics could not be removed consistently with the full and perfect security of our establishment in Church and State.

was also present at their public exhibition day on the 29th
July. Its uninterrupted success was a great gratification to him.

On the 10th July he attended the Annual General Meeting
of the subscribers to the School of Arts, when Mr. Cockburn
proposed that a suitable building for the School of Arts should
be got, and the erection of such a building would be the most
appropriate monument to the late Mr. Watt; ultimately it
was known by the title of Watt Institution and School of Arts,
and in 1886 by being united to the Heriot Trust, it is now
entitled the Heriot-Watt College.]

From Mr. Henry Hallam.

Killin, *July 28th*, 1829.

MY DEAR SIR,—You will perceive by the date of this, that
we are in the heart of the Highlands—our tour has been,
hitherto, as successful as we could reasonably hope, and far
more favoured by weather than when we parted last, there
was cause to anticipate. We have been to Glencoe and to
Staffa, the latter abundantly repays the trouble, if such it can
be called, of including it in a Highland tour, and indeed is
worth all the rest put together. You shall of course hear
more details when we meet. But my chief reason for troubling
you at present, is to request that you will inform me if you
have heard any recent tidings of Sir Walter Scott's grand-
child, as I should be very unwilling to intrude myself on his
hospitality at a moment of distress or deep anxiety. If,
therefore, it proves that the poor boy is near his end, I should
certainly make that a ground at once for declining the pleasure
which a visit to Abbotsford would afford me. Should you be
able to give me information about this, which is likely to
be known at Edinburgh, a line addressed to Post Office,
Callander, will find me. I beg to be remembered to Mrs.
Horner, as well as to your father and sister.

I am, my dear Sir, very truly,

HENRY HALLAM.

[Early in the month of August Mr. Leonard Horner's father caught a severe chill and became seriously ill.   He had before that been so well, and in such excellent spirits, and so affectionate to his six grand-children, who were staying in his house.   However he made a rally, and the medical attendants told Mr. L. Horner no longer to delay his return to London, as there was no immediate danger, and they were desirous that as his health was still very far from recovered, that he should undergo no painful excitement.   They also forbade him travelling by sea, so he hired a light landau, and with all his family (several of which were very young), posted to London, and sent all the luggage by sea.   This was a delightful journey travelling by Longtown and Langham, Carlisle, and Greta Bridge to York, where they spent a few days with Mrs. L. Horner's aunts, and then went on by Grantham, etc., to London.

But they had not been long home, before Mr. and Mrs. Horner were summoned back to Edinburgh on the death of his father, in his 79th year.]

To his Daughter.

Edinburgh, *November 2nd,* 1829.

MY DEAREST FRANCES,—* * * We heard to-day of the death of poor Christain Richardson.   It is sad to see one so amiable cut off at such an early age ; and the affliction to her poor parents must be great indeed.   Such events are of every day occurrence, and being so, are too apt to pass unnoticed.   It ought however, to be otherwise; and should make all who are young, beware of calculating with too much certainty upon life, and be better prepared for a change, should it please God so to order it.   Such reflections need not damp the cheerfulness and gaiety of youth, so benevolently implanted in our nature by our Maker ;   they will only check the needless folly and emptiness of thought of unwise people, and will temper lightness of heart with the sobriety of reason.   .   .   .   .   It being Sunday evening, Mr. Macbean proposed that his nieces should repeat some chapters of the Bible which they did with great accuracy; I should like very much to see my dear girls able to do the same. Besides if the selection is properly made, the most important lessons may be fixed very strongly in the mind by this method. I should like you and your younger sisters to try. The chapters I should wish you to learn are the following : the fifth and sixth of the Gospel of St. Matthew, the twelfth of the Epistle to the Romans, and the 13th of the first Epistle to the Corinthians.

In her last letter, your Mamma told Mary to pay you your first quarter's allowance. This is an important era in your life, when you are so far launched upon your own responsibility, and it is therefore, a fit occasion for me to offer you a little advice. I will begin, however, by saying that this advice is given from no want of confidence in you; I have the fullest reliance upon your earnest desire to do that which is right. You are started off with a good stock of clothes, so as to enable you to begin the first year with the important principle of spending less than your income. If you adhere to this, you never can be put to inconvenience. Pay everything ready-money. If you want anything, and have not the money to buy it, wait till you have; you will feel a little disappointment at first, but it will soon go off. Keep an accurate account of every thing you spend, and balance your money account every week upon a fixed day.

My kindest love to all your dear sisters and kindest regards to Miss Parker.

I am ever, my dearest Frances,

Your affectionate father,
LEONARD HORNER.

CHAPTER X.

1830—1831.

[This spring was a most trying one to Mr. Horner from the trouble some of the Professors at the London University gave him, showing every disregard to his instructions; and this worry preyed on his health. His endeavour at all times, was to be courteous and conciliatory, and the other Professors were indignant that he should meet with this treatment, and cordially upheld his authority.]

————

From Mr. Henry Hallam.

4th June, 1830.

MY DEAR SIR,—I congratulate you on your complete refutation of the conspirators against your credit. The path, thank God, is now clear, and I trust the Council will shew vigour. Their own characters are at stake. I have lent to Lord Auckland a draft of resolutions and a minute on your case, and on that of the professors; whether he and others will approve them, I know not; I shall concur in nothing which is not substantially the same; a full acquittal of you, and a firm assertion of our own authority. It is manifest to my judgment, that the removal of four or five persons is indispensable, and that we shall lose half our best professors by striving to keep the worst. Indeed, I shall not retain my seat in the Council, if feeble measures prevail.

I do not see why we might not get through the business to-morrow; but if timid men press for delay, I fear we must grant it for a few days; I am strongly for calling a general meeting *after* the] removal of the factions. There are imperative reasons for this, in the state of the finances, and the objections are grounded on no solid foundation.

Yours very truly,

H. H.

From Francis Jeffrey.

Edinburgh, 22nd July, 1830.

DEAR HORNER,—I am glad to hear that you are better, and that your University horizon looks less stormy; whatever may come of it, I think you have done quite right to stick by the ship, and not to give way to the first grumbling of mutiny. I expect to get to Loch Lomond next week, we shall be back at Craigcrook, however, about the 20th of August, and not much away, I think, for the rest of the season. I hope it will not be very late in it before we have the pleasure of seeing you and your ladies there. I scribble this, in the midst of many interruptions, and doubt whether it is intelligible, but you will guess, I daresay, at the meaning. With kindest remembrances to all your house.

Ever very faithfully yours,

F. JEFFREY.

[About the end of July, Dr. Clark* advised Mr. Horner to go abroad, and spend three or four weeks at Ems and Schwalbach. So accordingly he and Mrs. Horner and their two eldest daughters crossed over to Holland, and proceeded to the Hague from Rotterdam in a treckschuyt, a sort of boat on the canal. They stopped at Delph to see the monuments of William the Silent, and Von Tromp. At the Hague they visited the collection of pictures and a museum of Japanese wares, which at that period were unique. At Leyden they visited the Botanic Gardens, and saw a tree planted by Boerhave. At Amsterdam they were kindly received by Mr.

---

* Afterwards Sir James Clark, Bart.

Horner's old friends, the Vouts, whom he had known in 1814. They then proceeded by Utrecht and Arnheim, and joined the steamer on the Rhine at Nimeguen, and arrived at Ems on the 6th of August, when Mr. Horner went through a course of the waters. His four younger children were left at Dorking Surrey, under the care of their kind governess, Miss Parker.

While on this little tour, the news reached them of the "glorious three days" at Paris, which displaced Charles X. and placed Louis Phillipe on the throne of France. After spending a fortnight at Ems, they proceeded to Schwalbach, passing on the way the little town of Nassau, and the Castle of Baron von Stein. Here they met Mr. and Mrs. Labouchère of Holland, with their three daughters, and after a few days, as the weather was chilly, they proceeded to Wiesbaden, when they met a most charming Dutch lady, Countess von Randwyck and her son. After staying a few days there, the party went on to Frankfort, when they heard of the Belgians having risen against their Dutch King, and great disturbances. They returned home in September.

In the spring of 1831, Mr. Horner resigned the office of Warden to the London University. His health had suffered much from the vexatious annoyances that he met with, though he had also the cordial support of many valued friends, who however, felt he was wise to resign.

The family then went abroad to reside at Bonn, on the Rhine, which was at that time, a delightful and thoroughly German town. Mrs. Horner, with four of her daughters, left London in June, and went to Godesberg, a charming village about four miles from Bonn, and Mr. Horner with his other two daughters, remained a little longer in London to enable him to wind up his affairs, and he also went to Scotland for a short time. His eldest daughter soon after became engaged to Mr. Charles Lyell.*]

* Afterwards Sir Charles Lyell, Bart.

To His Daughter.

Edinburgh, 18*th July*, 1831.

MY DEAREST FRANCES,—Many happy returns of this day to
you my very dear girl.   I have this day bought a shawl for
you, which you will receive as the birthday offering of your
dear Mamma and myself.

I was agreeably surprised to-day by the receipt of your
Mamma's letter of the 10th, and more than surprised upon
opening it, to see the handwriting of Mr. Lyell.   I came
down in the James Watt (steamer) with his father, who told
me that he was in Hampshire.

He is a most active man, and most zealous geologist, for
while other men content themselves with turning over
dictionaries, and commentaries, to clear up their difficulties,
nothing satisfies him but to cross the sea, and break the
rocks of distant mountains, to clear up his geological doubts.
I shall not be surprised to hear that he is next heard of on
the top of the Schnee-Kopf.   As to his seeing me in Scotland,
that is not very likely, for in less than a fortnight, I hope to
be on my way back to London.   But it must have been a very
agreeable surprise to all of you, and you must have enjoyed
your distant walks with him very much.   I am very sorry that
he did not defer his visit until I was with you.

I wrote to your Mamma on Saturday, after my arrival here.
That day I went to Bonaly to dinner.   I went out with Maitland
and Ivory; a beautiful day, and everything looking in
luxuriance and order and freshness.   We got out just in time
for dinner, and we had the same scene to which you are no
stranger, in the little dining-room with the Baron* in his grey
coat, and a vast nosegay, the long row of dram glasses, Frank
bursting into the room after dinner, fresh from the bush, and
the tremulous reflection of the leaves of the willows on the

* The name given to Mr. Cockburn by his friends.

green wall behind the Baron, which used to delight Williams*
so much. . . . . Playfair is now busy in erecting Mr.
Dugald Stewart's monument on the Calton Hill, which
promises to be a very fine thing. . . . I went to the Chief
Baron's, and sat an hour with him and Mrs.¦Abercromby, in a
sociable way. They are about to see a great change in their
plans, for the Court of Exchequer in Scotland is to be abolished
immediately; the Chancellor brings in a bill for that purpose
in the House of Lords this evening. People here expect that
he will retire upon his full salary, but he does not expect it.
. . . This morning I breakfasted with blind Mr. Mackay
in Stafford Street, to meet Dr. Arnold, the head master of
Rugby, who is passing through Edinburgh¦ He is a very
agreeable man.

When we industriously set about our regular occupations,
for the fulfilment of a distinct useful plan, we shall find that
we shall have little time for gadding about. The chief object
of society will be the facilities it affords for the language;
which must be almost the exclusive object of attention. We
do not know that we may not be but a short time in the
country, and we must not miss the opportunity while we
have it.

And now my dearest Frances I must say farewell. Give
my affectionate love to your dearest Mamma and sisters, and
best regards to Miss Parker.

Your affectionate father,

LEONARD HORNER.

To His Eldest Daughter.
(At Godesberg).

14, Charlotte Square, Edinburgh, *July 20th*, 1831.

MY DEAREST, DEAREST MARY,—I wish that I had got your
letter, and Mr. Lyell's two or three hours sooner, that I

* Known as "Grecian Williams" from his beautiful paintings of Greece.

might have had leisure to collect my ideas on so very important a subject.

It is utterly impossible for me to write as I wish, for my heart is too full at the thoughts of the possibility of a separation from you, and I could give way to a flood of tears, were it not that it would be very selfish. I am bound to consider in this matter your happiness alone, and I leave everything to your own decision. As far as I have had an opportunity of knowing him, he appears to possess every essential quality to make you happy; sense, honour, and cultivation, and I have no reason to doubt his temper. . . On the subject of money I am not entitled to say more than that I must see that you will not be subjected to privations, and his expectations on the subject of present income, are such as I cannot object to. In short, my darling child, I leave everything to your excellent sense and feeling; you have the best possible adviser with you in your dear Mamma.

I am so anxious to be with you all, that I had made up my mind to give up the pleasure of going to Kinnordy, as I have so much to do; but now I will certainly go, that I may see those who possibly will be so nearly connected with you. I will write a few lines to Lyell by this post, directed to London.

Heaven bless my dearest Mary.

Your fondly affectionate father,

LEONARD HORNER.

————

To his Daughter.

Edinburgh, *July 26th,* 1831.

MY DEAREST MARY,—I wrote to Mr. Lyell at Kinnordy to tell him of my intention of being with them on the 30th and I then took occasion to explain my views to him. . . . . I received this morning an answer which is most kind and liberal, and dictated by the best feelings, and the assurance I felt before, that you will be made happy by this alliance, is

greatly strengthened by all the intercourse I have had with
Lyell's father. . . . . On Saturday, Maitland, George
Joseph Bell and Dr. Maclagan and I, went to Bonaly. At the
foot of the lane I got out to call at Woodville, the pretty villa
that belongs to Alison and his sons. I saw Margaret who
·gave me but a poor account of her father, who was too unwell
to see me. She shewed me their grounds which are very
pretty; she made many kind énquiries after you all. . . . .
James Gibson Craig came over from Riccarton. . . . .
Everything that I have seen and heard here, confirms me
more and more in the propriety in the step I have taken of
going abroad, to remain there while I am resting and considering at leisure what I ought to do. I have had many
combats to fight, but I think I have even convinced Cockburn
that I am right.

The Reform Bill goes on prosperously, but very slowly, in
the House of Commons, and I do not believe that it will be
finished there, before the middle of August. What is to be
its fate in the House of Lords, no one can tell. If it is not
knocked in the head very early, which I expect will be the case,
it will be a very long time of passing. I do not envy those
who have the management of it. I wait with some anxiety
for news from Holland. The Dutch have been disgracefully
abandoned by England in the settlement with Belgium. The
Belgians seem quite elated with their new king. This is the
anniversary of the Revolution in Paris. I daresay it will not
pass off very quietly, although things seem to be passing
more quietly in France as far as I know, but except what is
contained in the *Globe* newspaper, I see nothing of foreign
news, and I am not likely to hear any in Edinburgh, for no
one takes any interest in what is going on on the continent.
I have found no one to have thought at all about Holland
and Belgium.

I have heard from Susan and Kate to-day who appear to

continue to be much pleased with their visit to Ham.

My kindest love to dearest Mamma and all your sisters. If Lyell is still at Godesberg, give him my best regards. Kind regards to Miss Parker.

I am, my dearest Mary,

Your affectionate father,

LEONARD HORNER.

To Charles Lyell, Esq.

Godesberg, 16*th August,* 1831.

MY DEAR LYELL,—I arrived here with my two companions[*] all well yesterday. I found all my Dutch friends in Rotterdam in such good spirits, so little appearance of disturbânce, or any deviation from their ordinary course of life among the people of the town, and so decided an opinion among my best informed friends as to the absence of all danger, that I resolved to make the tour of Holland; but after sending the girls to bed, I went with Henry Labouchère about ten o'clock to the *Societat* (club) to hear the latest news from the Hague, expected by express that night, *viz.*, the communication from the King and the States-General, that day. Soon after ten the express arrived, and it was announced, that in consequence of the declaration of the allied powers sent to the King from London, that France had taken Belgium under its protection with the approbation of the other powers, that unless Holland immediately withdrew her armies within her own territories, France would advance with fifty thousand men, to support Belgium by land, and England would send Fleets to the Scheldt and Maas, that in conseqence he (the King) had ordered his armies not to advance.

This announcement created utter dismay among the audience, for they had just been put into high spirits, by fresh news of the

[*] His third and fourth daughters.

gallant behaviour of their troops. It immediately occurred to
me, that such a change would so entirely occupy people's minds
that my friends would be little at leisure to receive me, with
any comfort, and that warlike movements might take place
nearer Godesberg. There was not time to hesitate long, for if
I did not go by the steam boat of the following morning at six,
I must wait till Monday, and I at once resolved to set out and
run no risk of detention, and at half-past six we were off for
Nimeguen. I, had an opportunity of witnessing the good
behaviour of my girls on the occasion ; they had set their
hearts very much upon seeing Holland, and went to sleep
expecting to breakfast at Delft, and when I awoke them in
the morning, and told them of the change of plan, they
immediately turned to the bright side of the picture, that the
prospect of seeing dear Mamma presented. We got to
Nimeguen in time to see the town pretty tolerably well.
We landed at Cologne on Sunday afternoon and saw the Dom,
the skulls of St. Ursula's eleven thousand Virgins, and walked
upon the bridge for an hour. We started yesterday at six
from Cologne, got to Bonn at ten where I hired a carriage,
and drove up to this door at half-past eleven. What a place
this is ! It exceeds all my anticipations. What a view from
the Castle ! We walked there last night. What delicious
peacefulness, and *no University !* Most of the last twenty-
four hours has been occupied by me in conversations with Mrs.
Horner and Mary, about you. There is nothing of which I
do not think my dearest Mary deserving, but I assure you
with perfect sincerity, I am far more proud of having you for
a son-in-law than I should have been of any rank or fortune,
and I know full well she will be far more proud of the rank
you have obtained as a man of science, than she would have
been of any title however noble, of any fortune however great.
I look forward with the greatest comfort to the advantage I
shall derive from having ready access to a judgment in which

I am disposed to place the fullest confidence; having neither
son nor brother, I have felt the want of such an adviser,
although I have many very kind and intimate friends.

Very soon after this, you will be at Kinnordy; say everything that is kind from me to your father, mother, and
sisters.

Mamma is more disposed to look at the bright side of things
than she was, and I shall do what I can to increase that
disposition, This is Nora's birthday, and we are to have
some festivities, and I daresay *thirteen rosebuds*.*

I am, my dear Lyell,
Yours very faithfully,
LEONARD HORNER.

[After spending some months at Godesberg, beautifully
situated opposite the Seven Hills, Mr. Horner took an apartment in Bonn, where his family moved on the 6th September.

The society there consisted of two English families, the
Hare's† and Barton's; ‡ some of the old Nobles who still
lingered in the residence of the old Elector and Archbishop of
Cologne of past days; two regiments of Prussian soldiers,
(the Lancers and Hussars) and a number of professors. The
Bartons had a charming apartment, of which the ground-
floor was occupied by Professor Bleek and his family. It had
been the house of Niebuhr, who died there in the beginning
of this year. Of the Nobles, the most distinguished by riches
was a Westphalian, Baron Furstenberg and his pretty wife,
who lived on the Münster platz. Baron Böselager, and his
young wife, who was the daughter and heiress of the minister

* A custom the kind landlord practised in decorating the birthday plate
according to the years.

† Mrs. Hare Naylor, and her daughter, step-sister to Julius, and
Augustus Hare, became the second wife to the Rev. Fredk. Maurice.

‡ Mrs. General Barton with her sons, and a daughter who married the
Rev. Fredk. Maurice in 1836.

of the late Elector, in whose house they lived. It had the finest rooms in Bonn, square and lofty, a succession of them were hung with damask. Besides these, there were Count and Countess Beust, most intelligent and agreeable people. He was the head of the *Bergamt;* the late Austrian minister was his nephew. Among the Professors, August Wilhelm von Schlegel, the venerable Ernst Moritz Arndt, Dr. Mendelssohn, Professors Windischmann, Brandis, Harless, Nöggerath, Goldfuss, Treviranus, Sack, Nitsch and others—all men of wide reputation.

Mr. Horner enjoyed the respite from care, and the leisure gave him an opportunity of pursuing his favourite geology, and studying the German language and literature. He made a geological excursion in the autumn with Professor Mitscherlich of Berlin, who was passing some days at Godesberg.]

From Mr. Hallam.

Tunbridge Wells, 17*th October,* 1831.

My Dear Horner,—I was extremely glad to receive your letter a few days since in town, but most particularly on account of the intelligence it contained of Miss Horner's approaching marriage. The union seems to me all you could desire, in respect of the person she has chosen; a gentleman so highly esteemed for his private character, and so distinguished in his line of science, and above all, one with whom you have had such long intimacy. Though their income may not be very large, you justly observe that they are not dependent for happiness on things requiring much money; and I presume Mr. Lyell has some future expectations. Mrs. Hallam is as much pleased as myself, and sends her warmest congratulations to the bride elect, and to her mother.

We have been the whole summer at Hastings, till we moved hither about three weeks ago, and now think of taking up winter quarters in Wimpole Street very soon. Whether

we may remain there, at least regularly, after the present
winter, is not certain ; a desire to live more in the country,
which has been growing on me ever since I ceased to have a
permanent occupation, has much increased, both from some
domestic considerations (among which Mrs. Hallam's health,
unfit for much evening society, is one), and also from my
disgust at the prospects of public affairs, and so irreparable
an alienation, in political sentiments, from many of my
friends, that, however I may desire, as I strongly do, that it
should not lead to any personal estrangement of feeling, I
cannot but know by experience, that it is likely to bring
about a diminution of intercourse. It has always, however,
been a great problem with me, in what manner to accomplish
this object of rustication ; not altogether to try the experiment
of a mere country life, and yet unable to keep up an expensive
London house in conjunction with any other residence.   It is
therefore possible that these impediments may still detain me
in the course of life I have so long pursued.

Your own account of Bonn is very satisfactory ; if Europe
should remain in tolerable tranquillity, you seem to have
chosen a convenient, agreeable, and economical residence,
the use of such good libraries is invaluable, and I presume
you will not have failed to explore the recesses of the Sieben
Gebirge, and the volcanic regions of Andernach.

In one respect only, your letter disappoints me, that you
have not hinted a word about the state of political expectation
in Rhenish Prussia, a subject of no slight interest. I begin
to hope that the fever is a little subsiding on the continent,
while, heaven knows, we are in its paroxysm, at home.
The firmness of the French Ministry in preserving peace, will
I trust, have that reward which political firmness generally
obtains, and always merits.

I have little to say of the London University. Though not
injured by it as you have been, I heartily wish that I had

never been engaged in its concerns. The business is managed,
I hear, by Lord Sandon, Tooke, and one or two more, as a
select committee. The Medical School has opened, how
successfully I could not learn, the others will evidently fail
more and more, and I take it for granted the stoppage for
want of funds will occur not later than February. King's
College is also opened, but I have heard nothing of it. I have
left no space for politics, indeed I have no heart to say much,
the Peers have shewn two qualities, which mankind mu s
perforce admire, talents and courage; in the former all agree
in their superiority, during the debates on reform, over the
Commons, and I think the latter speaks for itself. Nevertheless, no one can doubt that the bill will be carried; but
whether by a sort of pretended compromise, for the Ministers
dare not, if they would, modify the bill in any important
degree, or, as would be more honourable for the opponent
Lords, by their withdrawing in a body, rather than perishing
in a fruitless resistance, I cannot say. There is not only
more self respect attached to their dying like men, but a
better chance of a joyful resurrection. Not that in this state
of the world I think hereditary privileges, when once gone,
have a very good chance of recovering, but it is impossible to
predict future events, and the best guessers are as often
wrong as right. We shall be always delighted to hear from
you.

Most truly yours,
H. Hallam.

*From Mr. Falck, the Dutch Ambassador in London, to
Mr. L. Horner at Bonn.*

London, *1st November*, 1831.

J'ai été fort sensible, mon cher Monsieur, à l'intention bien
veillante et amicale qui vous a engagée à m'ecrire
immediatement après nos succès du mois d'Avril. Ma reponse

a été retardée par une indisposition assez longue durant la-
quelle j'ai été forcé de me borner aux occupations strictement
necessaires, c'est à dire aux soins que reclament de M. de
Zuylen et de moi la negociation avec les cinq Cours.

Vous connaissez les 24 Articles qui sont le resultat de
cette negociation, et vous savez aussi que l'on n'en est pas
très content en Hollande. Il me semble qu'on a tort, et
que l'honneur national ayant été mis à couvert par la
campagne de dix jour, nous devions nous montrer faciles sur
l'arrangement de quelques inteérêts purement materiels. Le
point important à mon avis, c'est de pouvoir permettre à
nos bons Volontaires [de rentrer chez eux après dix ou douze
mois d'absence, et d'espérer ainsi une diminution notable
dans les dépenses excessives de notre état militaire. Si
nous refusons les Articles, je ne vois pas de terme à ces
sacrifices qui sont vraiment hors de proportion avec nos
resources, et de plus nous mettons de mauvais humeur
l'Angleterre, et les autres puissances qui jusqu'a présent
nous ont plus ou moins favorisés. J'espère que ces considerations finiront par faire quelque impression sur l'esprit
du Roi, mais en deliberant si longtemps, il donne beau jeu a
nos impatiens amis du "Times," qui ont dejà retrouvé pour
lui,l'usage de leurs anciennes épithets de *short sighted, obstinate,*
&c.

Je vous felicite de passer l'hiver hors de l'Angleterre. Il
vaut mieux se trouver à une certaine distance d'objets dont
l'aspect n'offre rien de consolant.

Ministerial ou Tory, tout homme éclairé, et vraiment ami
de son pays, me parait serieusement inquiet de la sourde
agitation qui regne dans les classes inferieures; et
malheureusement je ne puis me persuader que le remède
applicable au mal, soit, ou dans l'establissement d'une Garde
nationale recommandé par le "Morning Chronicle," et par des
familles plus influentes, ou dans la formation de sociétiés

politiques telles que Sir Frances Burdett a présidée hier
dans Lincoln's Inn Fields, et ou il a été de prime abord *out-
voted* et entrainé au delà de son bût par les orateurs des
Working Classes.   Dieu veuille que je me trompe, mais plus
d'un pays a l'air d'avancer a grands pas dans une carriéré de
dissolution et de trouble.

Veuillez me rappeller ainsi que M. de Zuylen de Nyevelt
aux souvenirs de vos Dames, et croyez je vous prie à ma
sincère amitie.

A. K. FALCK.

Vous ignorez peut-être encore que malgré la grande
incertitude sur le parti que prendra le Roi, vos compatriots
commencent à partager la bonne opinion, que vous avez
depuis longtemps, des funds publics de la Hollande.   Il s'est
debité ici depuis quinze jours, pour cinq ou six millions de
florins en certificats à 2½ per cent., et le prix de 42 a 43 qu'on
en donne, a sans doute réagé advantageusement sur la
Bourse d'Amsterdam.

Translation.

I felt very deeply, my dear Sir, the kind and friendly thoughtfulness
which induced you to write to me directly after our successes of last
April.   The delay in answering your letter was owing to a long illness,
when I was forced to do nothing that was not strictly necessary, that is
to say, only what was forced upon Mr. Zuylen de Nievelt and me, by the
negotiations with the five Courts.   You are aware of the 24 Articles that
have been the result of these negotiations, and you know also that they
have not been very satisfactory to Holland.   It seems to me that she is,
wrong, and that as her national honour has been maintained by that ten
days' campaign, we ought to yield with a good grace some points which
are purely material.   The most important object in my opinion, is to be
able to allow our good Volunteers to return to their homes after an
absence of ten or twelve months, and thus to diminish considerably our
enormous military expenses.

If we refuse these Articles I see no end to these sacrifices, which are
quite out of proportion to our resources, and besides that we shall
displease England and the other Powers, who have hitherto more or less
favoured us.   I hope that these considerations will, in the end, make
some impression on the King's mind, but by deliberating so long he lays
himself open to the sneers of our impatient friend, the *Times*, who has
already resumed its old epithets, calling him short-sighted, obstinate, &c.

I congratulate you on passing your winter away from England. It is of some advantage to find oneself at a certain distance from a state of things the aspect of which is anything but comfortable. Whatever party he may, belong to, whether to the Ministerial or Tory side, every man of enlightened views, every true lover of his country, must, it appears to me, be seriously uneasy at the agitation which is brooding among the lower classes ; and unfortunately I cannot persuade myself, that there can be a remedy for this bad state of things, either in the National Guard which the *Morning Chronicle* and some of the most influential men recommend, or by the formation of political societies, such as Sir Francis Burdett presided over yesterday in Lincoln's Inn Fields, and where he was directly out-voted, and driven far beyond the end he had in view, by orators of the working classes.

God grant that I may be mistaken, but more than one country seems to me to be advancing by great strides on a career of trouble and dissolution. Pray remember me and M. de Zuylen to your ladies, and I beg to believe in my sincere friendship.

A. K. FALCK.

P.S.—You are perhaps not yet aware that notwithstanding the great uncertainty, as to what part our King will take, your countrymen begin to share the good opinion which you have entertained so long, as regards the public funds of Holland. Within this fortnight they have placed five or six millions of florins at two and a half per cent. in our Funds, at the price forty-two and forty-three, which has doubtless reached advantageously on the Amsterdam Exchange.

CHAPTER XI.

1832—1833.

[Early in this year Mr. Horner was advised to be a candidate for the office of Treasurer of the Bank of Scotland. Though reluctant so soon to relinquish his happy leisure, he felt it his duty on account of his family, to go to Scotland, and make every inquiry about it; he finally gave up the idea from various prudential motives, much to the regret of his dear friend Mr. (Lord) Cockburn, who rejoiced at the prospect of his returning to Edinburgh.]

From Mr. Henry Cockburn to Mrs. Horner.

Edinburgh, 8*th March*, 1832.

MY DEAR MRS. HORNER,—Your worthy husband is glorious. A blacker, a dirtier man was never emptied out of a coach, than he on his first arrival out of the mail. But now that he is shaved and cleaned, and has bought a new vest, he looks lovely. His affair looks lovelier still. He has been treated with a degree of kindness and liberality, by everybody connected with the Bank, especially by such as are politically opposed to him, that would overwhelm and soften any modest man : what may be the result cannot be known certainly, but at present I have a confident hope of success. He some times affects to doubt his fitness ! (Ha ! ha ! ha !) and to fear the risks ! Risks ! Exactly those which disturbed the pensive ploughman when it occurred to him what would happen "*If the lift were to fa' and smoor the Laverocks ?* " that is in English, " *If the sky were to fall and crush the Larks.*" If he had been——

14th March.—A long pause————. Well, if he been in his soda water breakfast state, this delicacy would not have been wonderful, but in a man eating, as he does, at the least, four large meals of solid food every day, it is very odd. However, since I began, this modest fit has greatly abated, and he rather begins now to think that he is the very best man they could get. He is now dubbed and styled as "The Thesaurer," and we are all looking for large supplies of money from him, and so are our wives. I cannot tell the pleasure with which I anticipate the return of all here again, which experience should convince you is your city of refuge. And then Mary, she will be married after all according to the rites of a lawful and Christian church. Leonard is going to Kinnordy on Saturday, to return on Wednesday. We have not been at Bonaly yet. I go there for my vernal month next week: I was out on Saturday—the day (like all our days in this Arcadian winter), beautiful, and the grass teeming with crocuses; next day, being last Sunday, we went to Duddingstone, the minister* of which, has got his arm (his painting one too) eaten by a horrid dog. We came home over the hill. What a scene! Bonn! fiddlesticks! come here if you want views.

What an annoyance it will be if Leonard fails after all. But if he does, we shall console ourselves with the conviction that it is a low and dangerous place, and that there is no cholera at Bonn. We thought we had subdued the pest here, but are wrong. It is now very bad in the village called the Water of Leith, and is appearing all over the country. But in no one instance has it yet assailed any well clad, and well fed, lady or gentleman. We all send you our sincerest loves.

Ever and ever,

H. COCKBURN.

---

* The Rev. John Thomson, a very accomplished artist.

To Charles Lyell, Esq., Senior, of Kinnordy.

Bonn, *May 29th*, 1832.

As Charles has informed you a few days before I left London, I finally renounced being a candidate for the Treasurership of the Bank of Scotland. I am happy to think that my determination has been approved of by all my family, as well as by most of my best friends. A conversation with Charles fixed the resolution I was previously disposed to adopt. It was to the Solicitor-General* I wrote my final determination, and I have not had time to hear from him since, but I fear I must make up my mind to his thinking me to have acted wrong. I think I could reckon upon an unanimous vote in my favour at Kinnordy. I had a delightful walk in the Botanic Garden last night, looking upon the lovely Siebengebirge, and I felt like a bird that had escaped from the snares of the fowler.

My journey from London, thanks to the quarantine regulations in Holland, afforded me great pleasure. Had it not been for them, I should have passed over the profitless sea, and the comparatively dull flat country on the banks of the Meuse, Wahl, and Rhine between Rotterdam and Cologne. I saw the beautiful country between London and Dover from the top of an excellent coach; and well may that country surprise and delight a foreigner who sees England for the first time. I had a quick passage to Calais, with every advantage of weather. I had heard of there being a quarantine at Lille of five days, and to have a chance of avoiding it altogether or of having it abridged, I got, through the assistance of my friend Dr. Clark,† (King Leopold's English physician) despatches from the Belgian Embassy to carry to Brussels. At Calais 1 was advised to avoid Lille, and go by way

* Mr. Henry Cockburn.

† Afterwards Sir James Clark, Bart.

of Dunkirk, Bruges, and Ghent, an advice I was very
willing to take, having already seen the other road by Lille
and Tournay.   I went by the canal from Dunkirk by
way of Furnes and Nieuport to Bruges.   It is somewhat
tedious, but if one has books, and is not pushed for time,
there is great enjoyment in that mode of travelling.   The
canal is narrow, and the country is uninteresting, until you·
join the great canal from Bruges to Ostend, a magnificent
work, and then the country is very pretty.   I spent a couple of
hours in looking at the curious old town of Bruges and then
went on to Ghent by the night barge, where you can sleep as
comfortably as in a good inn.   I· passed ·a long morning in
seeing what was most curious in Ghent, and got in the afternoon to Brussels.   Next day I had a most agreeable journey,
I passed over the field of Waterloo, and saw all the great
points spoken of in the battle, including the observatory on
which Napoleon stood, then Quatre Bras, Ligny, and Fleurus,
all distinguished for the fights that preceded the great combat.
Namur is in a very striking situation at the confluence of the
Meuse and Sambre, and the drive along the banks of the
former river, by Huy to Liege, is one of the most interesting
I have seen in any country.   The same sort of country
continues to Verviers, passing the new watering place of Chaudes
Fontaines.   At Eupen in the Prussian Frontier, I was very
nearly detained on account of the quarantine regulations, but
happily escaped and got on to Aix-la-Chapelle.   Next day was
an easy day's journey to Cologne and Bonn.   From Bruges to
Verviers, Belgium is an uninterrupted tract of richly cultivated country with good houses, and a numerous well fed and
well clothed population.   I got different accounts as to the
state of the country, but I fancy there can be no doubt that it
is recovering fast from the stagnation occasioned by the
revolution.   Leopold is very popular everywhere, and they
are looking forward to his marriage with the daughter of

Louis Phillipe, as a very fortunate event. I found everywhere the greatest interests expressed upon the events now passing in England on the subject of the Reform Bill, for they consider that the return of the Duke of Wellington to power would be very injurious to their independence. Just on the contrary side is the feeling of the Dutch at Rotterdam, there was an illumination when the news arrived of Lord Grey having resigned. But I hope that before long the differences between Holland and Belgium will be settled, and that two countries so well calculated to contribute to each other's prosperity, will not be long enemies. I have this day been with the Burgomaster and Advocate, to ascertain if all our proceedings preparatory to the marriage* are strictly in accordance with the forms laid down by their laws here, and I do not find that we have neglected anything. I shall go to-morrow to the Protestant Minister to hear what the forms of the Church require, and I have no doubt that every-thing will be in order by the time of Charles coming here.

To Charles Lyell, Esq., of Kinnordy.

Bonn, *2nd July,* 1832.

CHARLES wrote to Mrs. Lyell three days ago, and I write thus soon again, in the hope that my letter may reach you on or before Thursday the 12th, the day now fixed for the wedding, because you will be happy to be with us in imagination.

Tuesday, 3rd July.—We project a picnic in the Seven Mountains to-morrow, if the weather proves good. We have had a very bad June, much rain, and often very cold. We have had very few true summer days yet. The last three have been much better, and the improvement comes at a most important time for this country, for another week of rain would have utterly destroyed all hope of a good vintage,

* Of his eldest daughter.

the vines being now in flower. If we get heat there will be
more grapes than for some years. Every other produce
that now covers the ground looks as well as possible. The
wheat has been in the ear about ten days, but it is still a
late season.

I packed up a box when I was in London, which arrived a
few days ago. It contained the copies of Rossetti's work.
I gave one to Professor Brandis, who is an accomplished
Italian scholar, and is engaged on a work upon Dante. He
has had only time to cut the leaves open, but is much
pleased with what he has seen, and says that as soon as his
lectures are over he will write a critique upon it, to make it
better known in Germany. When Niebuhr was Prussian
Ambassador at Rome, Brandis was his Secretary of Legation.

We called on Schlegel with the copy for him; we missed
him, but he called next day, and he begged I would convey to
you his best thanks' for the honour you have done him.
The copy for De Witte of Breslau, I expect to have an
opportunity of forwarding soon. I asked Charles to bring
with him the two volumes of Rossetti's Dante for Brandis, as he
had never heard of them, and they are now in his possession.
I shall lend *my* copy of Rossetti's last work to some other
persons here, who are likely to take an interest in it.

You have probably heard that Dr. Hartmann, of Blanken-
burg in the Hartz, is engaged in a translation of Charles's
work. I wrote to him of Charles coming here, and I had an
answer from him yesterday, sending the proof sheets of the
first two hundred pages of the first volume. He speaks with
great admiration of the work, and of the pleasure he has in
making the translation. I believe that the work will be
eagerly read in Germany, and that it will do here what it has
already done in England, *viz.*, excite an attention to the
subject among men of a high order of understanding, to
whom Geology, as previously studied, did not offer sufficient

attractions. Charles is amused by observing how odd it seems to see his own ideas set before him in a foreign tongue, and especially one of which the construction is so peculiar, as the German. It will be an excellent book to advance him in a knowledge of German. He has been very assiduous in his daily lesson since he came here. He has set his new sisters to work ⸲upon collecting the fluviatile, lacustrine, and land shells of this neighbourhood, and has built a drag net after the fashion of the one at Kinnordy.

Everything in the political state of this country is quiet, and the officers of the regiment here, do not expect that their thirst for a war with the French will be gratified. You will have noticed perhaps, the revolutionary movements in *Rhein Baiern*, that detached part of the kingdom of Bavaria, which lies nearest Deux Ponts. They were the work of some *têtes exaltées*, who planted trees of liberty in villages, and made some violent harangues, which had no other effect than to bring some additional troops into the country. There is nothing I have either heard or seen, which leads me to think that the subjects of Prussia are discontented, and except an extraordinary restriction of the liberty of the Press, which seems to me to be carried to an unwise degree of severity, even in reference to the country itself, and the forced military service, there does not seem any immediate pressure upon the people that should lead to discontent. I suspect, however, that much of the quiet is owing to the good personal character of the King. The despotism in the hands of a bad man, would not long be submitted to. I see in the English and French papers, that great assemblages of troops are taking place on the Rhine. There are certainly none here, and the officers tell me that there is no such thing elsewhere. So much for the accuracy of newspapers.

To his Daughter.

Bonn, 24*th July*, 1832.

MY DEAREST MARY,—On Saturday last, I was present at
Fritz Windischmann's* promotion. Schlegel was the great
actor and did the thing well. He read a Latin discourse upon
India, and Indian Literature, which lasted three-quarters of
an hour, but it interested me the whole time, and, notwithstanding his foreign pronunciation, I followed him quite
easily. The ceremony is rather imposing, and Fritz
conducted himself admirably well. Schlegel gave the Doctor's
Schmaus, and did me the honour to invite me to it. There
was nothing particularly striking about the dinner, although
the host had been so lately in Paris, but he made himself
very agreeable. When I came home I found Arthur Hallam,
who stayed two hours only, as he had to set off next morning
to England.

Thursday the 26*th.*—A very agreeable letter from Mallet.
He said they followed us in all our proceedings on the
morning of the 12th with the most earnest good wishes for
the young couple. At the moment I write, it is exactly a
fortnight since Mr. Sack was tying the knot, and it seems a
very long time. Mallet says that they have had a beautiful
summer, that fruit is in immense quantity, hay abundant,
and corn very promising. The Cholera increases very much,
and attacks people of all ranks, and the cases are as fatal as
at Paris. At Liverpool and Cambridge it is very violent.
The Assizes at York are put off in consequence of it. The
floor of the House of Commons is watered with a solution of
chloride of lime. The Scotch Reform Bill has received the
Royal assent; the Irish has gone through the House of
Commons. As far as they have yet gone, there appears
among the people, a general disposition to have gentlemen

* Who received a University honour.

and tried men for the New Parliament. But the Radicals are
at work, inculcating the importance of pledges from
candidates, to urge the abolition of tithes and corn laws;
Lushington and Macaulay and others have publicly protested
against candidates being called upon to give pledges.
Attwood and Parnell and Warburton are urging a free trade
in Banking, to the extent of every man issuing paper who
likes; so that until the Reformed Parliament has pounced
upon these matters, all property will be considered very
insecure, and thus a stagnation in the interchange of property
must continue much longer. John Romilly will certainly
come in for Bridport. The Chancellor has given his brother
James a sinecure in the court of Chancery of £2,000 a
year, vacant by the death of Lord Eldon's son. Lord Minto is
appointed Ambassador at Berlin; Lord Nugent, Chief
Commissioner of the Ionian Islands. I am quite bit with the
idea of going to study under Leonhard and Bronn at
Heidelberg for a week or ten days, and if I can make up my
mind to the expense, there is nothing else to prevent me.
It would do me much good. How very extraordinary that
Loess is? It has not been half attended to. I saw it to-day
in great abundance, and at a considerable elevation between
Muffendorf and the Rotherberg, and I found a bone in it
about three inches long, and one and a half broad, which
appears to me to be a portion of a rib of an animal as large as
a horse. Yesterday I examined a quarry of grauwacke at
Dollendorf beyond Kessenich, I found many impressions of
plants, and the whole rock of such a nature, as to associate it
with the old red sandstone or youngest grauwacke.

To-day, at two o'clock, we set off, a great party, to
Godesberg. After taking coffee we started for our walk.
Mamma, with a true spirit, suggested the Rotherberg; so
with the addition of three donkeys, we sallied forth, and a
famous scramble we had over fields, and reached the summit

of the crater.   After loitering there some time, we set out on
our return by way of Mehlem, where I had appointed the
carriages to take home the weary, and we were lighted all the
way by glow-worms and fire-flies, and I cannot end the day
better, than by having an *Unterhaltung* with my dear
Mary.

Ever your affectionate father,

LEONARD HORNER.

To Charles Lyell, Esq.

Bonn, 28*th August*, 1832.

MY DEAR CHARLES,—In my former letter I said that we had
given up our project of a visit to the Ahr valley.   Last week
I was arranging my visit of a few days to the *Siebengebirge*,
and I asked Mrs. Horner if she would undertake a walking
excursion, and go with me; she readily consented, but
Frances did not like to undertake it, so Kate had the next
offer and was delighted to go.   So, with nothing more than a
supply of clothes for all three, contained in one bag, we
started last Thursday, the 23rd, at seven in the morning, in a
one-horse open carriage.   I was dropped at *Rolandseck*, where
I had appointed Sassenberg* to meet me, but Mrs. Horner
and Kate went on to Remagen, where they were to breakfast,
and amuse themselves till I came up, which I did at one
o'clock.   I forgot to say that we took the opportunity of
conveying our maid Lina to see her friends, as Linz was
included in my plan, wishing to see the brown coal and the
basalt there.   We crossed from Remagen to Erpel, and the
ladies and Lina proceeded along the shore to Linz on foot,
with a boy carrying the bag, while I struck into the mountains
with Sassenberg, and after a tour of four hours joined my
party at Linz, and there we stopped.   We were at the very

---

* A guide who accompanied geologists.

mouth of the Ahr valley, and as I wished to leave a stronger
impression of the journey upon my companions, than there
was reason to expect, by their merely accompanying me in
my geological pursuits, I got another one-horse carriage at
Remagen, to meet us next morning at Kripp, opposite Linz.
We started at seven, and by the way mounted the Landskron,
and got to Ahrweiler to a late breakfast. We thought the
famed beauties of the Ahr valley by no means deserving of
the praise bestowed upon them, but our host at Ahrweiler
told us the beauties only begin there, and that we must go
higher up. We took his advice, and went up as high as
Altenahr, and we agreed that all we had heard fell short
of the beautiful scenery from Ahrweiler to Altenahr; above
all when the latter town comes first in sight, at the summit
of the pass. We got back to Ahrweiler at six, dined, and
arrived at Rolandseck soon after ten; a good day's work, and
well spent. We next morning crossed the Rhine to
Honnef, and I sent the ladies, under the guidance of a trusty
Frau, to the summit of the Hummerich, while I went another
way to visit other, and more distant summits, and afterwards
joined them.

We slept at Königswinter, and the next day I spent entirely
in the Sieben Gebirge, meeting my fellow-travellers at the foot
of the Oehlberg, returning to Königswinter in the afternoon.
Yesterday I spent in that part North East of the Königswinter
Thal. . . . . I have seen much that was new to me,
and much that interested me greatly. The Loess covers the
Basalt columns of Unkel, at an elevation of at least three
hundred feet above the Rhine, and here it contains large
calcareous concretions—compact limestone, with the land
shells enclosed. I found loess again at Orsberg, on the right
bank of the Rhine, exactly opposite to Unkel, covering the
brown coal beds, in which the frogs were found at an elevation
of not less than five hundred feet. I found brown coal beds

s

of lignite close to the summit of the Minderberg, a basaltic hummock which cannot be less than one thousand two hundred feet above the Rhine. At Obercassel, I found regular blue and yellow clay of the brown coal formation, containing peaty matter, inclosing the same shells as the loess—thus one part of the brown coal is younger than Nöggerath supposes. Sassenberg found out this place very lately, only mentioned it to me by chance, making no account of it; Nöggerath has not seen the place. Sassenberg says, it is the only place he ever found shells with the brown coal beds; I dug out a helix with my own hands, deep in the peaty matter. I have seen basalt dykes traversing the trachyte tuff in five different places, also traversing the trachyte itself. I have seen the grauwacke covering trachyte, and loess over both, a most capital section. With kindest love to my dear Mary.

Believe me always, very affectionately yours,

LEONARD HORNER.

To Charles Lyell.
(In Scotland.)

Bonn, *Saturday 29th September,* 1832.

MY DEAR CHARLES,—On the 17th Mitcherlich came here, on his way to pay a second visit to the Eifel. What we saw when we were there last year together, having interested him so much; particularly the volcanic bombs, upon which he has been making some experiments with a view to publication. You are aware that there are some of the bombs entirely composed of mica, and he says that he has fully satisfied himself by experiment, that they are in all probability the produce of altered fragments of slate, for he has obtained crystals of mica by melting grauwacke slate under particular circumstances. The third part of his system of chemistry (two are published), will be entirely devoted to what may be termed the chemistry of geology, and will be published in April. This is an almost unexplored field in which much is to be done, and coming out from him, cannot fail to excite attention. I proposed to him to undertake the translation of that part, as a separate geological treatise, and he is to send me the proof sheets, as they are thrown off, so that if I can find a publisher in London, the translation may appear as soon as the original. We spent a day in the Sieben Gebirge together, very agreeably, and profitably for me. Mitscherlich has read your book* in Hartmann's translation, and speaks in the highest terms of it. He says it is sure to excite great attention in Germany. Marcus, the bookseller here, told me, he was travelling with Nöggerath from Cologne, who was reading the work, and he said that nothing like it has appeared for a very long time. I found Meyer, the Professor of Anatomy, busy with it a few days ago. Nöggerath is preparing a notice of it for the Jena literary journal.

All unite in kindest love to yourself and Mary, and give our best regards to Mr. and Mrs. Lyell, and all your sisters

I am ever, my dear Charles,

Very affectionately yours,

LEONARD HORNER.

From Lord Minto.

Berlin, 30th October, 1832.

MY DEAR HORNER,—It was a great disappointment to me to have been obliged to pass through Bonn without seeing you. I had formed a little plan of passing a day there, and intended to have given you due notice of our approach ; but a five days quarantine at Spa, after an inevitable detention of two days at Brussels, left me no time to spare in reaching my post here, and I was very unwillingly obliged to renounce my visit to you.

* " The Principles of Geology,"

I can tell you little of our prospects here as yet; if we were people who depended much upon the amusement of general society, I do not think they would be brilliant, but I hope to find resources enough to make the time pass easily, independently of the political interests and occupation, which, since my arrival here, at least, has been sufficient to prevent me from finding the day too long.

I continue to hear excellent reports of our Scotch elections. Who would have believed it possible five years ago, that we should look to Scotland to fill the Whig benches of the House of Commons! or that our country gentlemen should have discovered that they are amenable to public opinion. By the way, talking of public opinion, I am a little curious to know what is the truth with regard to it in your neighbourhood; it is the fashion to represent you as full of loyalty and contentment. I am sure I shall be glad if this be true.

I receive most satisfactory accounts of the prospects of the new French ministry, which promises better than they had ventured to hope themselves.

Pray remember me very kindly to Mrs. Horner and

Believe me ever, most sincerely yours,

MINTO.

P.S.—Mr. Horner left Bonn early in February, 1833, to visit the Charles Lyell's in London.

To his Daughter.

Church Row, Hampstead, 21*st March*, 1833.

MY DEAREST KATE,—I have just retired to my bed room, the front room in Mr. Mallet's house, with a comfortable fire, and sit down with great satisfaction to have a little talk with you. First let me tell you about my kind host and hostess. They are both well, saving a slight cold which has confined Mr. Mallet for some days; and the boys are perfectly

well. Charles and Mary went yesterday to his Aunt, Mrs.
Heathcote's, and I mean to stay here till Saturday. We
were quite alone to-day which was particularly agreeable
to me. . . . . I must now
give you an account of what I have been about since
my last despatch. On Saturday, I dined at Mr.
Labouchère's*. We had the Mallets, Sir Robert† and Lady
Inglis, Francis Baring‡ and his wife, and some others.
Young Henry Labouchère of Rotterdam sat next to me. The
house is a very splendid one, beautifully furnished, and he
has some good pictures. Charles called for me about ten
o'clock, and we went to one of the Soirées of the Duke of
Sussex, as President of the Royal Society, at Kensington
Palace. There were a great many people of note there, but the
person who most interested me, and whom I had been long
desirous of meeting, was Talleyrand. I heard him talk with
great vivacity for a quarter of an hour, and was close enough
to see the minute expressions of his very remarkable
countenance. It was very interesting to see in this way a
person who has played so great a part in the public affairs of
Europe. The Duke's apartments are very spacious, most
luxuriously and handsomely fitted up, but not with any show.
In almost every one of the several rooms there are shelves full
of books, and there is one room, the library, with books all
round. I spoke to Mr. Sydney Smith there, and introduced
Charles to him. He called on Mary on Monday. Sunday I
dined with the Murchisons, and went to Mr. Greenough's Soireé.
On Monday I dined at Mr. Marshall's in Upper Grosvenor Street,
and among many persons whom I did not know, there was

---

* Peter Cæsar Labouchere, a partner in the great mercantile house
  of Hope, father of Lord Taunton, he died 1839.

† Sir Robert Inglis, Bart. M.P., born 1786. Died 1855.

‡ Francis Baring, nephew of Mr. Labouchère, created Baron
  Northbrook in 1866

one whom I had met before, Mr. Brunel, the engineer of the
Thames Tunnel, a clever and interesting person. Last night
I went for a quarter of an hour to Sir George Phillips's. The
Edward Romillys were there, and Sir John Dalrymple. This
morning I called on the Murrays in Duke Street, Westminster.
They are very well, and little Willie is much delighted with
London. I then went to call on Mrs. Marcet, who has had a
cold, and sat an hour with her, she was very agreeable. The
church clock whose sound is so familiar to me, is just striking.*
Many a time has it struck as I was returning from visiting
Miss Lloyd.

Your affectionate father,

LEONARD HORNER.

[While Mr. Horner was in England, an enquiry had been
made in Parliament as to the treatment of children in
factories. From that a commission arose, and Mr. Horner
and others were appointed to enquire into the employment of
children in factories, on the 19th of April, 1833. He
entered upon this office on the 30th of April, and with his
colleague, Mr. Woolryche, a late surgeon in the army, and a
very agreeable man, he visited the factories of Gloucestershire,
Somersetshire, &c.

The first report of the Commissioners was dated June,
1833, and the " Factory Act " founded upon that report, was
passed on the 29th of August, 1833.]

To his fifth Daughter.

Gloucester, 11th May, 1833.

MY DEAREST LEONORA,—I assure you I am heartily tired of
being so long separated from you all; it is fourteen weeks since
I left Bonn. I wrote to your Mamma from Worcester, where
we visited, as Commissioners, the porcelain works of Flight
and Barr. We saw a most beautiful service for the King,

* Where he was married.

which costs above six guineas each plate; the expense arising
from the painting. There was also a very pretty service ordered
for the Pacha of Egypt. We walked into the Cathedral, there
is a very fine piece of sculpture by Roubilliac, a monument to
Dr. Hough, Bishop of Worcester, who died in 1743, at ninety-
three years of age, after having been fifty-three years a
Bishop; a very learned and virtuous prelate, who distinguished himself while President of Magdalen College,
Oxford, by his resistance to the arbitrary proceedings of
James II. The monument is by the same sculptor who
executed the celebrated statue of Sir Isaac Newton, at
Cambridge, and fine as that is, I am not sure that I do not
prefer the figure of the Bishop. There is nothing in modern
sculpture which I have seen equal to it, and a monument by
Chantrey which is near it, is a sad falling off from
Roubilliac. In the choir of the Cathedral King John lies
buried, his tomb was accidentally discovered about thirty
years ago, and the robes in which he had been buried were
then tolerably fresh; the teeth were quite fast in the jaws.

We ordered lampreys for dinner, which are found in the
Severn, near Worcester, as I had never tasted that delicacy
so celebrated in ancient and modern *gourmandise*. Yesterday
we went to Great Malvern to breakfast. I never saw a more
lovely country than that between Worcester and Malvern,
and it is just now in the highest beauty. It is mostly
pasture land, with numberless orchards. We passed one
which was the largest and finest orchard I ever saw. I think
there could not be less than three hundred pear trees, all of
great size, and they are just now white with blossom; there
are beautiful cottages all the way. I was delighted to see
Malvern again after so long an absence, having the most
pleasing impression of our stay there in 1810. It is much
altered, being, I should think, four times as large as it was
then. I looked for and found the cottage in which we lived,

and remembered the gravel walk, where dearest little Mary
ran about with pads on her knees. I walked up to St. Ann's
well, and to the summit of the Worcestershire Beacon, where
I looked down upon Herefordshire in all its beauty. I took a
hammer with me, which I borrowed from a blacksmith, and
recognised the old granite. Perhaps you do not know that I
published an account of the Mineralogy of the Malvern Hills.
You will find it in the first volume of the Geological
transactions. It was my first geological paper. We had a
beautiful drive from Malvern by Upton on Severn, and
Tewkesbury to Gloucester, where we arrived yesterday afternoon.

Stroud, Sunday 12*th May.*—We walked through Gloucester
yesterday, and were much struck with the singular sight of a
large Basin with great ships in the midst of a town in the
centre of England. A large canal, called the Berkley Ship
canal, brings ships from the Severn to Gloucester, which has
thus of late years become a considerable seaport, though so
far from the sea. We saw in the Basin or Harbour, a large
three-masted vessel that had brought from Memel, in the
Baltic, a cargo of wood. We left Gloucester after dinner,
and had a very pretty drive to this place, nine miles distant.
This morning I had a walk of five miles before breakfast,
up one of the valleys, and a more lovely country I never saw.
It is chiefly pasture land, with a quantity of fine trees, the
ground diversified by hill and dale, and dotted all over with
white cottages. I never was in a country presenting such an
appearance of a numerous well conditioned peasantry, and all
the people I see, seem well fed and are remarkably well
dressed, not a barefooted child, although the weather is now
warm.

Wednesday 15*th May.*—Still at Stroud, where we shall be
for three days more, for this is the great centre of the
clothing district of Gloucestershire, and there are above a

hundred mills in the neighbourhood. Here is made the
finest cloth that is used, and I believe *now* the finest and
best of any made in any part of the world; formerly they
made better in the Netherlands and France, but the English
manufacturer now fully equals anything that is done abroad,
and even surpasses it. We have seen great factories, where
eight hundred people are employed, and where, within the
same building, every process is carried on, from the unpacking
of the bale of wool, to the sending out the cloth ready for the
tailor, some as fine as thirty shillings a yard. The superfine
broad-cloths are made of Saxony wool; they used to employ
Spanish wool, but the merino sheep of Spain were taken to
Saxony, Silesia, Bohemia, and different parts of Germany,
and taken so much better care of, that their fleeces were
much improved, and the wool became much finer than
anything to be had in Spain. The great manufacturers here
go over every year to Germany to buy their wool. There are
great fairs held at Leipzig, Breslau, Berlin, and even so far
as Pesth, in the months of June, July and August, where the
wool is brought in bags, and the Princes who are proprietors
of flocks, often attend those fairs to sell their wool. I saw a
manufacturer to-day, who attends the Breslau Fair, and they
generally buy the wool from the same flock of sheep for
several years in succession. He told me that he had this day
got a letter from his agent, saying that he had bought eight
tons, the produce of *one* flock; now as one sheep does not
yield more than three pounds, there must have been about
six thousand sheep in the flock, and he tells me that there
are sometimes flocks of twenty thousand. The last house we
visited belongs to a manufacturer, who treated us with great
hospitality, and told us that the house we were then sitting
in, had been occupied by Friar Bacon. There were many
religious houses in this valley. There was a large convent
of nuns at Merch in Hampton, who came over from

Normandy. Many of the manufacturers we have seen are extremely opulent, and so far from there being any necessity for new laws to protect the children employed in factories here, any interference would be sure to do harm; I never in my life saw a greater number of rosy cheeked, well fed, well clothed children, than I have seen working in the mills here. No where are bare feet to be seen, and in a factory of several hundred people which we saw to-day, there were not more than three who could not read, and these were old people. Much as I enjoy this country, my thoughts are every day turned to the Rhine, and I am like a schoolboy counting the days when I am to see you all again. God grant that I may have that blessing in five or six weeks at most, and of finding you all in perfect health. We shall probably leave Stroud on Sunday, we go to Dursley, Wootton-Underedge, Bath and Frome, which last place will occupy us some time, as it is the great seat of the Somersetshire clothing district, as Stroud is of the Gloucestershire.

Give my kindest love to your dear Mamma and all your sisters, and I include Sophia and Elizabeth* as my daughters at present. Mr. Thomas Lyell, I presume, has left you. I hope your drawing is going on well, and that you will all be very active in the fine weather in out-door sketching. Never mind fine views or striking objects, practise is what you want.

God bless you my dearest Leonora.

Your affectionate father,
LEONARD HORNER.

———
To his Daughter.

Bath, 22nd May, 1833.

MY DEAREST JOANNA.—Another month will, I trust, bring me very near you. I was looking at the new moon last night,

---

* Sisters of Mr. Charles Lyell, who had come to pay a visit to Mr. Horner's family at Bonn.

and said to myself, "I hope I shall see the next new moon reflected on the Rhine." I am seeing a beautiful country just now, and many interesting objects, but all these are of far inferior interest to me, than those I have at Bonn. We were detained at Stroud till Tuesday forenoon; we proceeded by a most beautiful country to the banks of the Severn at Sharpness Point, near Berkeley, where the great Ship Canal that goes to Gloucester, joins the Severn. The country we passed through is all pasture land, and is that in which the Gloucester Cheese is made, which one sees in every part of England and Scotland. We passed by several dairy-farms, and as it was about four o'clock, the cows were collected in the farm-yards to be milked. Some farmers have a hundred cows. Sharpness Point is a beautiful wooded Promontory, overhanging the Severn; we did not see it in full beauty, for the water in the river was low; you must know that the tide rises in the Severn as far as sixteen miles above Gloucester, and the difference at Sharpness Point, between low and high water, is thirty feet. At Chepstow on the river Wye, which flows into the Severn, the difference is sixty feet. From Sharpness Point we went to the little town of Berkeley, which is rendered famous by having been the residence of Dr. Jenner, who found out the cow-pox to be a preventative against the small-pox, and he found this out by living in a country where there are a great many cows, for it was a curious thing that the people who were much occupied with milking, never had the small-pox; and this clever man, did not let that observation pass, without reflecting upon it, and thus was enabled to bestow a blessing on the human race. The cows of Berkeley have saved the lives of thousands in the remotest regions of the Globe. But Berkeley is also renowned for its castle, and *that* we did not fail to see. It was built about seven hundred years ago, and is kept up in the best order, and in the old-fashioned style by the Proprietor, Lord Seagrave,

the descendant of the Earls of Berkeley. It is in a beautiful
situation. After going round the castle, we were brought to
a small room in what is called the keep, and in that room
King Edward the Second was put to death in the year 1327.
After a hasty dinner at the Inn, we drove over in a most
beautiful evening to Dursley, five miles off, where we slept.
It is a small town, formerly supporting many cloth
manufactories, but now a good deal gone to decay. It is most
beautifully situated in a valley surrounded with wooded hills.
Next morning we drove up another pretty vale to the village
of Uley, where we stayed some hours visiting some large mills ;
we returned to Dursley to dinner, and in the evening had a
charming drive to the little town of Wotton-Underedge.
There we passed the greater part of yesterday, as there are
several mills, which we had to visit, and we got to Bath last
night. You are now, I hope, enjoying fine walks every day.
I see a great many horse-chestnuts in full blossom, and they
always put me in mind of the Popplesdorf Alley. I hope
Mr.* and Mrs. Sterling and their children have arrived safe.
How happy Mrs. Barton and Anna must be with them. I
hope I shall certainly be in Bonn, before Charles and Mary
come, for I should sadly grieve to miss one hour of them there.
I trust too that they will stay over the anniversary of their
wedding day, and if it please God that we are all well together,
we shall have a merry meeting on the occasion.

God bless you my dearest Joanna, my kind love to Mamma,
your sisters, Sophy and Elizabeth.

Your affectionate Father,

LEONARD HORNER.

[The summer passed away most pleasantly, both from the
intercourse with their kind friends at Bonn, and from visits
of various people passing through. Mr. and Mrs. John
Sterling were staying with her mother, Mrs. Barton ; Mr.

* The Rev. John Sterling.

Julius Hare, Mr. Bowstead (afterwards Bishop of Lichfield), Captain and Mrs. Basil Hall, Mr. Dawson Turner, and his daughters, Lord Cole (afterwards Lord Enniskillen), an ardent geologist, and his friend Mr. Egerton, Sir Francis Head, all spent a little time at Bonn, also the eminent botanist, Mr. Robert Brown, whom they then first made the acquaintance of, and afterwards became very intimate with.

In the autumn Mr. Horner's two eldest daughters at home, went to Scotland, with the Miss Lyells, who had been staying the summer at Bonn, with Mr. Horner's family.

On the 29th October, there was a grand day at Bonn, on the occasion of a visit from the Crown Prince, the pupil of Niebuhr, and afterwards King of Prussia. He resided during his stay there at Baron von Böselager's; there were illuminations and a ball given in his honour.]

From Captain Basil Hall.

Geneva, 10*th* August, 1833.

DEAR MRS. HORNER.—You wished to be informed of Mrs. Hall's confinement. The young stranger made his appearance the day before yesterday, and if all goes right, we think we may be starting for Italy in about six weeks hence, after making some excursions to the most interesting places in this neighbourhood. My two little girls are much amused with their little brother, who already begins to supersede the dolls and wheelbarrows in their estimation, and as he advances in life, it is to be hoped he will not lose interest in their eyes. We were much disappointed at the Lyell's not coming this way. I wrote to him lately, and addressed my letter to Bonn, (23rd July), and suspect he must have gone before my letter reached, as I asked him to let me know what his address was to be. I have since then finished reading his book, and I am rather desirous of writing about it while the subject is fresh in my thought. It is a most admirable, and truly a wonderful performance; the first book by far on the subject;

or indeed almost upon any scientific subject, which I know of.

Pray remember us most kindly to Mr. Horner, if he be with you, and believe me,

Ever most sincerely yours,

BASIL HALL.

[As before stated, the Factory Act was passed on the 29th August, and four Inspectors were appointed, with Superintendants under them. The British Islands were mapped out into four great divisions. The cotton and woollen districts of Yorkshire, Lancashire, and the ' immediate neighbourhood forming the first: the eastern and southern counties of England the second: some parts of the west of England, nearly the whole of Wales, and the southern part of Ireland constituting the third: the northern half of Ireland, the whole of Scotland, and the four northern counties of England, the fourth; surgeons were appointed to grant the certificates required for the children. Each district was placed under one Inspector. The gentleman who was Inspector of Factories in Scotland and the north of Ireland, either died or resigned the office in the end of October, or beginning of November 1833. Francis Jeffrey was then Lord Advocate for Scotland, and the appointment was placed by Lord Melbourne under his patronage, and he offered it to Mr. Horner. Lord Melbourne was Home Secretary at the time, and the appointment was in his gift. By desire of Lord Melbourne, Mr. Young wrote to offer the situation to Mr. Horner, requesting an immediate reply; and having accepted the office, he started for London on the 8th November. The happy Bonn life was at an end, to the great regret of the younger members of the family. Mrs. Horner, assisted by her three younger daughters, had to settle all the business of the move, as furniture had to be disposed of, and books, &c. packed, and they left Bonn early in December, having bid goodbye to many valued friends. They remained in London during the winter.]

To Mrs. Horner.

Brussels, 10*th November*, 1833.

MY DEAREST ANNE,—Here I am safely arrived. . . . .
I spent my time in the *Eilwagen* meditating on the sudden
change in our affairs, on the prospects it opened up, and on
our residence during the last two years and a half; and all
the beauties of the Rhine, and the country around Bonn,
were fairer than they had ever been, and I passed in review
the many kindnesses we have received from the excellent
people it was our good fortune to live amongst. . . . . .
I hope you, and my dear Kate and Nora had a pleasant party
at the Countess Beust's. Do not forget to give my kindest
remembrances to every one of my friends whom it was not
possible for me to call upon. . . . . . God bless you
dearest love.

To his Daughter.

Belfast, 29*th November*, 1833.

MY DEAREST MARY,—You are, I suppose, according to your
plan at this time in Edinburgh. . . . . . I have broke
ground in my new vocation very auspiciously, as far as a
good reception from the mill owners goes. They naturally
dislike the Act, like any other interference, but they say that
as they were to have one, that which has been passed is very
little open to objection, and they see no difficulty in carrying
it into effect. They have all said that they will cordially
co-operate in all the provisions which concern the education
of the children, and indeed in any other. Nothing could be
kinder than the way they have received me, and one of them,
Mr. Lepper, has placed his little carriage at my disposal, and
I am to dine with him to-day. I have seen a good deal of my
excellent and benevolent friend, Mr. Joseph Stevenson, whose
cotton mills, about a mile from Belfast, are a model for

everything that concerns the well-being of the work-people,
bodily and mentally. He is Secretary of the Belfast Academy,
and a great ally of Spring Rice. I see the means of doing
much good to the children, especially as regards their
education. Mrs. Bateson* is remarkably well, more like sixty
than eighty. I dined with her on Wednesday, and met Sir
Robert and Lady Bateson, Dr. MacDonnel, and Miss
Montgomery. Sir Robert and Lady Bateson are very well,
and asked kindly after you. Yesterday I dined with Dr.
MacDonnel, who, I am glad to say, is in great vigour. We
spoke much about geology, and about Charles. He says
"the Principles" is a *grand book*. He is no ordinary judge.
I sent a letter to your dear Mamma to Aix la Chapelle, and
another to Brussels. I trust they will have fine weather. . .
Thank you, my dear Mary and Charles, for your wish that we
should be long with you. What a treat it will be to Mamma
and ·you, mutually to be so long together, if we stay in
London all the spring, or rather if she stays—and the more
I think of it, the more I am inclined to consider that
arrangement the best.

I cannot yet tell what leisure I shall have for the *Penny
Cyclopædia*. I would gladly continue the articles, but I fear
that it might be a burthen to be tied down to a particular
day, in my circumstances. I will certainly go on with the
Penny Magazine, and I have a great wish to execute the
article on volcanos. But for the first five or six months, I
shall have to turn my mind almost exclusively to my business,
and I see such a means of usefulness in it, that I like the
prospect very much. The absence of all control, except that
of the Secretary of State, and the independence in action
which I feel, is a prodigious advantage.

I have a few notes for Charles. He speaks somewhere about
tigers being found near the Caspian. When I was writing

* Mrs. Horner's Aunt.

Ararat I found old Tournefort describing how he and his companions, when they ascended that mountain, threw themselves on the ground, *close by the region of eternal snow,* to escape some tigers which they saw passing by. What a pity it was that Cockburn could not accept the invitation to Kinnordy, and how agreeable it is to hear that Mr. Lyell would have enjoyed his company. I hope another opportunity will offer. I am rejoiced to hear from Charles that dear Fan and Sue are favourites.

Your affectionate father,

LEONARD HORNER.

CHAPTER XII.

1834—1835.

To his Daughters,
(Who were staying at Kinnordy.)

Arbroath, 14*th April*, 1834.

MY DEAREST FRANCES AND SUSAN,—You are anxious to
know how I have been received by the mill-owners, who, in
December, were very much disinclined to the Act, and who
have hitherto been quite unused to any legislative interference.
I have been agreeably disappointed, wherever I have been
from Kirkcaldy to this place, I have met with the kindest
reception, and found a very prevalent disposition to do the
best they can to fulfil the intention of the legislature.   I
have put such things in train, that I think before many
months go about, there will be new schools erected here and
in Dundee, by the mill-owners, especially for the factory
children.   I have no fear but that in a short time the
greater proportion of the mill-owners in my district, will
view the Act, not only without dislike, but will even admit
it to be in many respects highly beneficial.   I shall go this
day week to Perth, and then to Blairgowrie, and hope to get
to Cupar Angus on the 23rd, and I shall write in time for
Mr. Lyell to know the day I expect to be at Glamis.

After spending a day or ·two at Kinnordy, I shall propose
our setting out together for Edinburgh.   We are to have the
Murray's house till Parliament rises. . . . . . . The
ball at Dundee was not the only gaiety I entered into there.
Mr. Horsley, the Episcopal Minister, asked me to dine with
him last Thursday, to meet Kean the actor, and to go
to the play in the evening when he was to play.   I

went, and was much pleased. Kean is only twenty-
three, and a very agreeable, well bred, unassuming man.
The play was Massinger's " New Way to pay old Debts," and
he performed the part of Sir Giles Overreach admirably.
It was a curious feeling to pass from the quiet young man at
dinner, to the storming, purse-proud, mad Baronet in an
hour, and the change of dress was so great, that I should not
have recognized him anywhere. We all went back to Mr.
Horsley's to supper, and talked over the play with Kean. I
was so much taken with the youth, that I called upon
him next day, and I went to his Benefit that evening, when
he played Hamlet, and excellently well.

Having some spare time on Friday, I had a geological
walk along the shore with Mr. Macvicar, a young clergyman,
very intelligent and enthusiastic in science, a great friend of
Dr. Fleming, the naturalist. I had a delightful walk on
Sunday along the sea-cliffs.

God bless you, my dearest girls,

Your affectionate father,

LEONARD HORNER.

To his Daughter.

Durham, 11*th May*, 1834.

MY DEAREST MARY,—I left Carlisle by the mail on
Wednesday, a little beyond Brampton is Naworth Castle,
a fine baronial mansion belonging to Lord Carlisle. I
despatched some business in Newcastle that afternoon and
next morning went to Darlington. My first visit was to the
mill of Mr. Pease, a Quaker, and the father of the Member—
nothing could be in more beautiful order than his mills,
which are for spinning woollen yarns and worsted. On
Friday I went to Stockton, a neat town with a remarkably
broad street. Yesterday morning I drove to Barnard Castle

to breakfast, a very pretty country by the side of the Tees.
I saw Rokeby and Raby on either side in the distance.
Barnard Castle is a poor looking place, but there is a fine
view of the Castle, which must have been at one time a grand
building; it stands on the edge of a high rock overhanging the
Tees. I got back to Darlington in time for a coach which
brought me here in time to see two factories. I shall go on
this afternoon to Newcastle and spend to-morrow in visiting
four mills. . . . . . I expect to be in Edinburgh on
Tuesday evening. I see by the *Times* that Jeffrey is to be
the Judge, and Murray the Lord Advocate. There will be all
the bustle of an election, both in Edinburgh and Leith,
immediately. My amusements lately have been reading
Goethe's "Faust," Scott's "Minstrelsy of the Scottish Border,"
and Herder's "Leben." The life of Herder is very instructive
and entertaining, and gives me a greater desire than before,
to become acquainted with the writings of that accomplished,
amiable, and highly moral person. Every morning that I am
not obliged to work or travel before breakfast, I go on with
my grammatical study of German, and feel that I am
making some progress in a more exact acquaintance with the
language; I feel no repugnance to grammar as some people
have, the study of a language is always very entertaining to
me.

God bless you, my very dear pet,

Your very affectionate father,

LEONARD HORNER.

———

To his Daughter.

Kilmarnock, 24th June, 1834.

MY DEAREST MARY,—On Friday morning Mr. Smith, from
Deanston, called upon me in his carriage before breakfast,
to take me to Ballindallock Mill, near the town of Balfour.
We drove through a pretty country, and I saw a mill in

better order than almost any I had previously visited. If it
had been in North Holland, there could not have been greater
nicety, for there was scarcely a nail's head that was not
polished bright.   The workers were all particularly tidy.
From Balfour we drove to Balgair Moor, where there was
a great cattle fair, which interested Mr. Smith, who is a keen
farmer ; but as I am not skilled in horn cattle, I was more
occupied with the fine view of the Highland mountains than
anything else.

On Saturday morning the Kippen coach brought me into
Glasgow by way of the Campsie Hills. I got my letters, saw
some persons on business, and set off in the Kilmarnock
coach at four o'clock. I had a letter from Colonel Stewart* to
say that I should find him at home, and that his gig should
be waiting my arrival at Kilmarnock. It was so, and I had a
pleasant evening's drive to Catrine, where I arrived about
nine.   Catrine is a small property about two miles from the
village of Mauchlin, which belonged to Dugald Stewart's
father, and he used to spend a great part of his youth there, and
afterwards when he was a Professor.   Colonel Stewart has
built a house, the old one being now in the midst of a village,
and laid out some grounds very prettily.   The house is only
two stories, both low, and covers much surface ; there is one
large room, the library, which is very pleasant, and is
surrounded on all sides by his father's books,† and many of his
own acquisition. He was very hospitable, and is a clever
and agreeable man, with a vast fund of information on many
subjects.   He seems to lead a very active life, both as a
country gentleman, and in literary occupations.

About forty years ago Mr. Alexander of Ballochmyle, a
rich man from India, and a neighbouring proprietor, built a
large cotton mill, which thirty-five years ago came into the

* Son of Professor Dugald Stewart.

† Left by Colonel Stewart to the United Service Club, in Pall Mall.

possession of Kirkman Finlay's father, and other partners; the most active of whom was a Mr. Buchanan, who had been bred under Sir Richard Arkwright. They prospered greatly, and Catrine, which was only a farm-house with a mill and blacksmith's shop adjoining, has risen into a village containing a population of two thousand seven hundred souls. On Sunday I went to the chapel, built by the proprietors of the works, where there was a large and most respectable congregation, and I afterwards called on Mr. Buchanan, who has a very handsome house prettily situated in ornamented grounds, half a mile from the works. He received me with great kindness, and walked with me to Colonel Stewart's, and got him to dine with them next day, as he wished me to dine with them. Yesterday morning I breakfasted with them, and spent the whole morning in my examination of the works; I find them in the highest order, and every arrangement that could be desired for the comfort and welfare of the people. A great many of the workers have houses belonging to themselves, which they have purchased out of their savings; I visited several of them, and was much struck by their extraordinary neatness, cleanliness, and handsome furniture. One of the mechanics (those who make and mend the machinery) a man of forty years of age, who is now working for twenty-five shillings a week, lives in a house which cost him two hundred pounds, and which he paid out of his savings. It is of two stories, with a nice garden, full of fruit trees and vegetables, and the rooms are most handsomely furnished. I do not think there can be a happier population, and it is quite delightful to see Mr. Buchanan among his people, who seem to look up to him as a father. He has himself a very nice family, and his whole establishment is in perfectly good taste. In going through the works, I observed in the bleaching department, some mineral incrustations, which throw a good deal of light on the formation of natural productions, and I have collected materials which will be capable of being wrought up into an interesting paper for the Geological Society.

On Sunday Colonel Stewart drove me through some very pretty country by Auchinleck, the seat of the late Sir Alexander Boswell, and Barskimming, the seat of Lord Glenbee. They are situated on the rivers Lugar and Ayr, which flow through precipitous banks of red sandstone, similar to Roslin, but finer.

I propose sailing for Belfast to-morrow. It gave me great pleasure to hear of the very pleasant letter you have had from your dear Charles,* when you write give him my kindest love ; tell him I travel with him in thought. I look into my newspaper for the advertising of the third Edition.

Adieu, my very dear Mary,

Your affectionate father,

Leonard Horner.

[On the 7th September, 1834, the British Association for the Advancement of Science, held its third meeting at Edinburgh. Mr. and Mrs. Horner entertained a large party at breakfast each day, during the six days of the meeting. The Charles Lyells arrived to stay with them, and among the company at these breakfasts, were Messrs. Arago, Agassiz, Sedgwick, Whewell, Peacock, Vernon Harcourt, Buckland, Greenough, Roget, Moll from Neuchâtel, Sir Charles Lemon, Mr. Fergusson of Raith, Sir. Thomas Dick Laudor, Dr. afterwards Sir William Hooker, Mr. Phillip Duncan of Oxford, Mr. William Hamilton and others. The meeting was a brilliant one, and at its conclusion Mr. Phillip Duncan of Oxford wrote the following lines in the album of Mr. Horner's youngest daughter, a little girl of eleven years old.

* Charles Lyell, then absent in Sweden.

" The feast of reason and the soul,
The flow is o'er, we've drained the bow
The guests are gone, and all that light
Which lately shone so pure and bright,
Is quenched, that light shall never here,
To many eyes again appear.—
Those planets in their annual course,
Obedient to their guiding force,
Shall shed their light on other lands,
Attracting other eyes and hands
To tell how brightly tho' they're gone
They once upon this city shone.
But ah ! how grievous 'tis to think,
On each returning year some link,
Shall drop from off this friendly chain.
Till not a single part remain."

" Et bien, Monsieur Arago !
Nous avons eu un farrago—
De science et philosophie
Je pense que ce mot suffit.
Il a été pour ainsi dire
Un scientifique délire."

On the 15th September, a dinner was given to Lord Grey
in Edinburgh, and Mr. Horner was one of the stewards;
2,600 people were at this dinner, in a temporary building
erected for the occasion. A procession of trades went to meet
Lord Grey on his entrance into the town, and thousands of
people assembled and received him with great cheers.

Mr. Horner went to Glasgow the following week, and in
November he went to London to a meeting of the Factory
Inspectors.

During the winter and spring of 1835, Mr. Horner left his
family every few weeks on tours of inspection, in different
parts of Scotland, the North of England, and Ireland.]

————

To his daughter.

Hilltown, Co. Down, *5th June*, 1835.

MY DEAREST KATHARINE,—My last letter was from Belfast.
I dined that day at Dr. McDonnel's and met a pleasant party.

The person whom I was most struck with was Dr. Crolly, the
Catholic Bishop of Down and Connor, who has been recently
elected Archbishop of Armagh and Primate of Ireland, but
has not yet been confirmed by the Pope. He is a sensible
and agreeable man, and I never heard more truly Christian
sentiments uttered by any minister of religion of any sect.
All his conduct corresponds to those sentiments, and were all
the different priests, whether Episcopalians, Presbyterians, or
Catholics, like him, "peace and good will towards men"
would be triumphant in Ireland.

Mrs. Bateson came to tea. On Wednesday morning I set
out in a car, and drove to Larne. It was a beautiful day, and
as the thorn hedges and the furze are covered with blossom,
the air was filled with their fragrance. I got to Mr. Barkley's
house about four and stayed all night. We had a small
party of very pleasant men; and among them Mr. Ward,
the Episcopal clergyman,, one quite worthy of being named
with Bishop Crolly. I started in a car at seven and drove to
Carrick-fergus to breakfast. It was a warm lovely morning,
the sea views beautiful, and the hedges in such luxuriance of
blossom that it was like driving in a flower garden.

I started this morning soon after eight, in a car, and had a
pleasant drive of twenty-two miles to Killileagh, one of the
prettiest spots in the country. There is a very fine old
castle, in good preservation (and inhabited), of which there
are authentic records up to the time of Henry II., how far
back it goes beyond that period no one can tell. I came
to Killileagh to visit a cotton factory, belonging to a Mr.
Martin, which I found in the highest order. As there was no
conveyance to be had in the village, the only car having
gone early in the morning to Belfast, Mr. Martin drove me
over in his own car to Downpatrick, the county town, and a
very neat one, containing a great many very handsome
houses. I visited the churchyard to look at the grave and

monument of St. Patrick. From thence I hired a car to take
me to Castle Wellan, where there is a mill belonging to a
Mr. Murland, who most kindly pressed me to stay to dinner,
which I did. His mill in the best order. He spins Flanders
flax into yarn fine enough to make cambric. Such a thing
was not supposed a few years ago to be possible. I mean
spinning flax by machinery so very fine. God bless you, my
very dear Kate.

Your very affectionate father,
LEONARD HORNER.

[In the autumn Mr. Horner went abroad for some weeks,
first to Paris, and then to Bonn, where a meeting of
naturalists was held, and where he was pleased to renew his
acquaintance with old friends.]

To his Wife.

Hotel de Bruxelles, Paris, 25th August, 1835.

MY DEAREST ANNE,—Here I am, safe arrived in this
wonderful place. We had a glorious passage of two hours
and a quarter, and landed at Calais at eleven, and after a
comfortable dinner at Dessin's, with a fellow passenger, I set
off in a coach and arrived at Boulogne soon after eight. Next
morning I walked out to survey the town, first to the beach
to look at the rocks, and I then walked about the town,
which I should have liked to have spent some hours upon. It
was low water, and there are very fine sands. There were, I
daresay, fifty bathing machines of one sort and another,
and it was a busy scene; a vast number of English, but
no one that I knew—even by sight. All the arrangements
for the bathers seem to be most excellent, and I never saw
such comfortable machines. I then walked about the town,
which is a gay, pretty place, with many excellent houses and
shops, and remarkably clean. There is a very excellent

museum, which I visited. It is open to all on Sunday, and
there were a great many of the common people (fishermen
and their wives and daughters), in peculiar, but neat and
tidy costumes. It is very remarkable, that notwithstanding
the frequent intercourse of the two countries, what a striking
transition there is from what you see at Dover, to the
appearance of everything at Calais. The Diligence, from
Calais, came up at one o'clock, and in half an hour we
started. I had one companion in the coupé, who proved
agreeable. He was an oldish gentleman, a M. Pannifax, a
native of Carlsruhe, but settled in Paris. He had lived in
London, and afterwards in Havre. He knew the Haldimands,
Marcets, and Achards, and John Rapp, intimately. We
dined at half-past five, at Montreuil, and arrived at Abbeville
at eleven, where we had café-au-lait. Here my companion
left me, and I had the coupé to myself the rest of the
journey. We arrived at Paris at five o'clock, and I have got
a comfortable room in this hotel. To lose no time I
determined to go to some spectâcle, and the company at the
table d'hôte agreed in recommending the Opera Français,
and a M. Bouvieux, a gentleman from Havre, kindly offered
to go with me. The piece was " Le Dieu et la Bayadère,"
an opera by Auber, and some of the music was beautiful, but
the great charm was Taglioni's dancing, which was wonderful.
It is a most extraordinary combination of strength and
grace. There was a great deal of good dancing besides, and
there is a great contrast in the grace and light movements of
the whole Corps de ballet, to what I have seen in London.
We had afterwards the overture to *Guillaume Tell*, and the
masquerade scenes in *Gustave*, which were splendid. After
the Opera, M. Bouvieux carried me to some remarkable
places in the neighbourhood, arcades of beautiful shops, and
to the Palais Royal. We took an ice, and a glass of curaçoa,
and then came to our hotel.

God bless you, dearest Anne, my affectionate love to my
five darlings, how I wish you and my girls were here!

Your very affectionate,

LEONARD HORNER.

———

Paris, 26th August, 1835.

I drove to the Geological Society, and found the clerk, who
told me the address of M. Boué, near the Palais du
Luxembourg. I found his house under repair, and after
knocking at several doors at last came to the kitchen, where
I found Boué and his wife, both in *deshabille*, sitting at a
table with bread, butter, greengages, peaches and wine, and
the servant *washing* by their side. Boué recognized me
at once; though he had not seen me for twenty years. He
gave me information about many of the geologists, and I
heard with regret that many are absent from Paris. He
was very kind in offering to be of use to me here. I then
drove along the Quay Voltaire and Quay d'Orsay, passing by
the house where Voltaire lived, and crossed the Pont des
Invalides to the Champs Elysées and drove as far as the Arc
de Triomphe. On my return to the hotel, having learned
that Professor Berzelius of Stockholm was still here, I
called upon him; he was very agreeable and is going to Bonn
and is to live in the house of Professor Bischof, where I am
to be. I dined at the table d'hôte, and then went to the
Theatre de Vaudeville, and was much amused, particularly
with a piece called *l'Habit ne fait pas l'homme.* We have no
idea of comic acting in England of this light sort; what
strikes one most is the perfectly natural manner of the
actors; you can hardly persuade yourself that it is not real
life, and the costumes and decorations are so excellent and
appropriate.

After breakfast I got into a cabriolet and drove to the
Rue St. Dominique to enquire about M. Brogniart and

learned he had set out for the Rhine last night. I drove to
the Observatory to call on M. Arago, and found him busy
with his lecture, and so could not see me. He sent out a
message that he would call upon me *bien sur*.

I afterwards went to the Jardin des Plantes, and saw
M. d'Orbigny, one of the Conservators of the Museum of
Geology, and settled to call on M. Cordier on Saturday, who
will show me the volcanic collections from Auvergne, which
have a great resemblance to the rocks of *Siebengebirge*. I
walked to the part of the gardens where the animals are
kept, and saw the giraffe to great perfection; it is a
wonderfully formed creature, but not graceful, it is far too
short between front and rear for its height.

August 27th.—After finishing the letter which I sent you
this day, my dearest Anne, I went to deliver my letter to Mr.
Benjamin Delessert. He is a banker of large possessions,
and one of the most influential men in Paris among all
classes, and. is a Member of the Chamber of Deputies. He
received me with great politeness and kindness. He gave me
an order of admittance to the Chamber to-day, and procured
for me admission to the annual meeting of the *Academie
Francaise*, and invited me to dine with him at his country
seat, Passy, on Saturday. His town house is a very fine one,
in the Rue Montmartre, I walked along the Boulevard until I
came to the Rue de la Paix, crossed through it to the Place
Vendôme, where I stood for some time admiring the triumphal
column, with Napoleon on the top of it, and I looked at the
corner of the Place, where I was in 1816, with poor Frank in
the house of M. Monterond. I then crossed the Pont Neuf, to
the Palais des Quatre Nations (built by Cardinal Mazarin),
where the sittings of the Institute are held, there I got my
ticket and heard that the doors would be opened in less than
an hour. In the interval I called on M. Deshayes, who lives
very near in the Rue des Marcies, off the Rue des Petits

Augustins. He was at home, and I found him very pleasant.
I am to call on. him to-morrow to see some parts of his
collection. By the favour of M. Delessert, I had got admission
to the principal place for strangers. The Institute consists of
a re-union of the old Academies, that of Belles Lettres or
L'Academie Française, L'Academie des Inscriptions et
l'Academie des Sciences. It was the annual meeting for
the distribution of prizes for the first of these. It is a handsome circular hall, lighted by a cupola, and in four niches are
marble statues of Descartes, Sully, Bossuet, and Fenelon.
There were a great many ladies. Mr. Pentland came in and
sat close by me, and said that he and M. Arago had just been
calling upon me. M. Tisset was President, and M. Villemain,
who is *Secretaire Perpetual de l'Academie,* read the announcement of the prizes. One was given in favour of a poetical
Eloge of Cuvier, by a M. Bignan. The hall is not good for
hearing, but I caught some good lines. I saw Humboldt.
The meeting lasted two hours and a half, and as soon as it
was over, I hastened to the Chamber of Deputies. The place
of meeting is a semicircle, of great dimensions, fitted up very
commodiously for the deputies, with a double gallery for
strangers. The effect is very good. One of the Vice-Presidents
was in the chair, but no one had any dress or costume of any
sort except the door keepers and attendants. I saw some of
the ministers : Guizot, Thiers, and Human, but I heard no
famous speaker. I was there an hour, and in that time four
deputies mounted the tribune, but only one read. One spoke
from short notes, two spoke extempore, and they were the
best. The subject was the new laws for repressing the licence
of the press, it excites much interest at present. The Radicals
are furious against the proposals, but I know too little of the
circumstances to judge of the merits of the case. When two
of the speakers were in the tribune whom they were not
disposed to listen to, there was a far greater noise than I ever

heard in the House of Commons, and the President had
constant recourse to his bell to repress it.. As I walked along
the Quai Voltaire, I purchased, for four francs, good lithograph
portraits of Cuvier, Humboldt, Arago, and Vauquelin—came
home, and found a most kind note from Lady Keith regretting
she was not at home yesterday when I called, and asking me to
call again to-morrow at an earlier hour. Count Flahault is gone
a shooting excursion with the Duc d'Orleans, and regrets that
he will not be back before I go. I dined at a restaurant in the
Palais Royal, and then went to the Theatre Français, and saw
the "Femmes Savantes" of Molière, and the "Monsieur
Tucaret" of Le Sage, (author of "Gil Blas") most admirably
acted, and as I got near the stage, I heard well and have not had
a greater treat for a long time. In the former the costume of
Louis XIV. was worn, and I could fancy myself back to the
time of Molière, when his troop were playing before Le Grand
Monarque in the Theatre of the Palais Royal. In "M. Tucaret"
they wore the costume of Louis Quinze. At ten o'clock I went
to call on M. Deshayes, who shewed me some of the remarkable
things in his collection, and I found his conversation very
instructive. I should be very glad, if it was in my power, to
put myself under his instruction for six months. I stayed
with him about two hours, and proceeded in a cabriolet along
the Quai Voltaire to call on Lady Keith, Rue de la Charte,
Champs Elysees, and found her at home. I passed through a
suite of most beautiful apartments, to that in which she
received me, which was a sort of drawing-room. She was
very kind, and chatted most agreeably for three quarters of an
hour, and told me a great deal of what is going on, at this
moment of great excitement in the politics of France. She
gave me her card of admission to the gallery of the Corps-
Diplomatique in the Chamber of Deputies, and I went there,
but did not stay long as there was a deputy in the Tribune
reading a discourse, and I could not hear a word. From

thence I went through the Garden of the Tuileries to the
Bibliotheque du Roi.

August 29th. Immediately after breakfast I set out in
a cabriolet for the Jardin des Plantes to visit M. Cordier.
I went along the quays on the left bank of the river, passing
the famous Place de Grêve, where so much blood was shed
during the Revolution. I found M. Cordier (who is Professor
of Geology at the Museum *d'Histoire Naturelle*), very obliging,
and I stayed five hours with him, studying the volcanic rocks
of Auvergne, and getting a great deal of useful information
from him. From thence I went to call on Mr. Pentland in the
Rue de l'Université, in that part of the Faubourg St. Germain
opposite to the Louvre. Then home to dress for dinner at
M. Benjamin Delessert's, at Passy, where I went in a
cabriolet. I was the second who arrived, and found his sister-
in-law, Mrs. Francois Delessert, in the drawing-room, who
was soon joined by her sister-in-law. It is a house of great
extent, situated on a height, commanding a beautiful view of
Paris, with a garden sloping towards the Seine. M. Delessert
is a considerable botanist, and his garden is full of rare and
beautiful plants, palms, orange trees and many of the greenhouse inhabitants placed out of doors. Mdme. Delessert
shewed me the garden, and a house built after the model of a
Swiss Chalet, the lower part of which is a dairy. M. Delessert
did not arrive till past seven, having been detained at the
Chamber of Deputies ; the discussion on the law about the
press, &c., having been brought to a conclusion, and the
session having terminated. The measures proposed by the
Ministers have been carried by a majority of 73. There were
ten of the Deputies besides at dinner, and M. Seguier, formerly
President of the Chamber of Peers. I sat next the Deputy
from Cognac, a native of Ireland and a wine merchant. He
discoursed much about brandy, as was very appropriate, but
I would rather he had talked more of politics. I talked much

with Mdme. Delessert about the Marcets, Prévosts, Romillys, &c., with all of whom they are very intimate. It was a very splendid dinner in all respects. I set out to walk home, but had not proceeded far, when a coachman driving a fiacre offered to take me as far as the Palais Royal for six sous. I accepted his offer, and was set down very near my hotel.

To his Wife.

Paris, 1st September, 1835.—I set off yesterday morning early, and was at the Jardin des Plantes by half past nine. I found M. d'Orbigny, who led me through the various rooms containing the collections of natural history. They are very fine, and I could spend a month in the mineralogical and geological departments with great pleasure and advantage.

I was particularly anxious to see the organic remains, or rather the fossil bones, of the neighbourhood of Paris, which have rendered Cuvier's name so illustrious, and made such an epoch in the history of geology. I had an appointment with M. Cordier at the Jardin des Plantes at 11, and I passed two hours with him, going over his collection from the *Cantal* and got much useful information. I then proceeded to the post office, and had just time to get back to the Palais des Beaux Arts, on the Quai opposite to the Louvre, where the meetings of the Institute are held, and to which Arago proposed to introduce me. I found him in his Bureau, (he is *Secretaire Perpetuel de l'Academie des Sciences,* and succeeded Cuvier) and he was very friendly. He introduced me to Baron Humboldt, who received me with great cordiality, and with whom I talked a good deal. Professor Carus of Dresden, a distinguished physiologist, was with him, I had made his acquaintance at Cordier's on Saturday. Arago put us both under Humboldt's charge, and he took us to see a remarkable specimen containing marks of the footsteps of a quadruped

U

upon a stone, a fact of the highest interest in geology; it
was found at Hildburghausen, and Humboldt read an account
of it to the Institute. The room of meeting is very handsome,
and very impressive, from the portraits of the great men
which surround it, who have been members of the Academy;
there is a marble statue of Racine, and another of Molière
opposite; then I saw the portraits of Boileau, Voltaire,
Rousseau and many others. Round the room there are two
rows of benches for strangers, within that, an oblong circle,
around which the members of the Institute sit, and in the
centre of it, the President and the Secretaries, and in front of
them a chair and table for the person who reads: *within* the
circle, and at right angles to the President's chair, are six
chairs, destined for the more distinguished visitors. On one
of these seats I was (most unworthily) placed by Humboldt,
by the side of Dr. Carus. The President was Baron Charles
Dupin, to whom I was afterwards introduced, as also to M.
Gay Lussac, the celebrated chemist. It is a day to be remembered all my life. I then went home to dress for dinner. 'I
arrived at M. de Flahault's at half past six. As soon as I
went up to Lady Keith, she said, "I asked you to a family
dinner, but I thought you would like to see some of the more
remarkable people, so I asked the Maréchal Gérard, and M.
Dupin, the President of the Chamber of Deputiès." He was
not less kind, and as agreeable as ever. It was a most delightful party. We walked for a short time in a very pretty
garden looking upon the Champs Elysées; and sat down at a
round table: Monsieur and Madame de Flahault, the Maréchal
Gérard, the President Dupin, M. Ramboteau, (the Préfet of
the Department de la Seine), Mr. Acton (the Secretary of the
English Ambassy). *Beautiful* dinner in all the arrangements,
and of course, the best style of cookery and wines, but what
was infinitely superior, the conversation of so many able
persons; we went together into the drawing-room, formed a

circle, and the conversation went on uninterruptedly for two
hours, the sole topic was the all-engrossing subject of the new
laws, and both the President and Maréchal Gèrard strongly
deprecated and deplored the extent to which the restrictions
have been carried, as did M. Flahault and the Prèfet. Some
new powers were necessary to repress the insolent licence of
the press, and some of the theatres, but the moderate Whigs
(for there is an analogous party here) consider that things
have been carried to an excess. I had got a billet d'entrée for
the Chamber of Peers to-day, and so I proceeded to the Palais
du Luxembourg, got a good place in the gallery, and saw and
heard well. Like the Chamber of Deputies, it is a half circle,
the President sitting at an elevated table in the centre. Baron
Pasquier, the President of the Chamber, was present, and I
had a full view of all the ministers, the Duc de Broglie,
Messieurs Guizot, Thiers, Marechal, Maison, &c. All the
peers are dressed in a uniform coat. It has not a good effect,
because they wear it with their ordinary morning trousers and
waistcoats of all colours. There was much more decorum
observed, much more silence kept, than in the Chamber of
Deputies. The subject was the new law for regulating jury
trials, and there was some good speaking on both sides; the
best was M. Persil, the Minister of Justice. I heard and saw
the vote called; the names of the peers are called over, and
they advance and put balls into two urns, according as they
are to vote—the measure was carried in favour of ministers
by a vast majority. Count Flahault said a few words. He
beckoned to me, and sent one of the huissiers to me, to
conduct me to him, and he then pointed out to me many of
the most remarkable persons, introducing me to the Count de
Presnil, with whom I had some conversation. I saw the Duc
d'Orleans, a very gentlemanlike young man.

To his Wife.

Luxembourg, *6th September*, 1835.

My Dearest Anne,—My last letter to you was from Paris; I was seated in the coupé for Metz at 7 o'clock, the road lies by Pantin and Bondy to Meaux, where I looked upon the Cathedral and Episcopal Palace of the celebrated Bossuet. There was nothing pretty in the road until we came to Ferté sons Jouarre, which lies in a valley well clothed with wood and vineyards. We had a long hill to walk up from Ferté, and the day was intensely hot. The next town of importance was Château Thierry on the Marne; at the entrance of the town is a wide space where there is a marble statue of Lafontaine, who was born there. At half past seven we reached Metz. The country I passed through is very various in aspect; part of it is celebrated in modern history, for at Valmy, General Kellerman defeated the Prussian army in September, 1792, commanded by the Duke of Brunswick, a defeat which had immense influence on the history of the French Revolution. At Verdun I looked with some interest on a place which had for ten years been the abode of the English *detenus*.

I got up early this morning to walk about the town. It is the largest garrison in the north of France, and the fortifications are very perfect. The place was never taken though often besieged. What pleased me most was the Cathedral. It is about the size of that of Lichfield, and of richly ornamented and beautifully light Gothic architecture; the interior is a more perfect style than almost any I have seen, and what adds to the beauty of it, there is no white-washing or colouring, there is the clean pure stone, and it is of a soft grey colour. It was seven in the morning and service was going on, the Sacrament was administering and the chapel was full; groups were kneeling on the ground in various parts of the great nave, so that the effect as a picture was striking. The outside of the Cathedral is deformed with shops, and with a Grecian front in the worst taste, added in the time of Louis XV. The vicinity of Germany began here to show itself by several sign-posts in German, it rejoiced me to see them again. I set out at ten o'clock in a diligence. When I came to Luxembourg I was well pleased that I had changed the route I had been advised to take, for it well deserves being seen, as a fortress of prodigious strength, and situated in a most remarkable manner on the summit of precipitous rocks, on one side which form the side of a deep ravine. It is wholly garrisoned by Prussians at present, as you know it is not yet settled whether it is to belong to Belgium, or Holland, or to Germany. There is fully as much German spoken as French, and the names of the streets are written up in both languages. There was a fair this afternoon which I went to see. At dinner I made the acquaintance of a worthy lady of Coblentz, an itinerant dealer in the finer kinds of lithographic prints, and she tempted me to buy two, the " Madonna di San Sisto " of Raffael, and the " Last Supper " of Leonardo da Vinci.

Treves, September 7th.—At six this morning I left Luxembourg, and was more and more struck with the singularity and beauty of the situation. It is well worth going a good way to see. I was very fortunate in my companion in the coupé, the Comte de St. Priest, *Pair de France,* who left Paris a week ago to make a tour on the Rhine, and specially to see the Moselle. He proved soon a gentleman of great accomplish-ment and high bred manners, a friend of the Duc de Broglie, and let out from time to time that he had been living in the best society wherever he had been, which has extended to St. Petersburg. He has also been in England. The diligence proceeded slowly, for there were many hills, and we had only two horses, but I did not mind, for we passed through a beau-

tiful country, and I had sometimes most agreeable conversation
with my companion, and I had a good book, an interesting
geological work of Elie de Beaumont on Auvergne. We
entered the Prussian territory at a small village at the confluence of the Sure with the Moselle, and there I saw again
the black and white posts which I had not seen so long. I
had no difficulty with the custom-house officers. From that
point the drive all the way to Treves is through a beautiful
rich alluvial plain about a mile wide, through which the
Moselle flows, bounded on either side by low hills covered with
vineyards, the red rock breaking out at intervals, and white
houses intermingled with the green; the distant view of Treves
(or Trier) is also good.

We arrived at one, and stopped at an excellent house, der
Trierscher Hof. The table d'hôte was just ready, so I and my
fellow traveller sat down, the greater part of the company were
Prussian officers. I could not stand the length of a regular
German dinner, but as soon as I got enough to satisfy my
hunger, set off to the post office, and to a bookseller's to get a
guide, and learn what was to be seen. It was a most lovely
afternoon, not too hot. I have just returned, and have seen
all the most remarkable things. I began with the great thing
of all, the Roman Gate,* which is a wonderfully fine thing.
Though called the Roman Gate, its date is much earlier than
the conquest of the country by the Romans, and it is con-
jectured to have been the work of a colony of Etrurians.
Next, I went to the Amphitheatre, about three-quarters of a
mile off, and that too is a most interesting remain. The
Prussian Government have cleared away a great deal of the
rubbish with which many parts were concealed. I do not, of
course, attempt any description of such places. To have
looked for the first time in my life on such works was a very
interesting moment. I then ascended a hill through vine-

* Porta Nigra.

yards, and had a very fine view of the town and vale, which is truly splendid. I returned to the town and walked through the principal streets, reminded at every step, by a thousand things, of Bonn. The houses, the people, the soldiers in precisely the same uniform, the boys going to school, the rolls in the baker's window, &c., &c.

On my arrival I heard, to my dismay, that the boat for Coblentz goes only twice a week; to wait two days was terrible, to go by the Eil Wagon, and so not see a mile of the Moselle, was also terrible. I set to work; and the end has been, that Comte St. Priest and I have hired a boat between us, and accompanied by his servant, we start to-morrow morning at five; we shall be two days and a half going down, sleeping at Berncastel, and Kocheim. The weather is beautiful. We have an awning for the boat, and Philip Nagel is an ex_ perienced *Schiffer.*

My love to my darlings. I shall write next from Bonn, I hope. God bless you.

Yours affectionately,

L. H.

To his Wife.

Bonn, 10*th September*, 1835.

My Dearest Anne,—I put a letter into the Post office at Treves for you on Monday evening, and the next morning Count de St. Priest, his servant Francois, and I set out for our boat, and at six we were afloat on the Moselle, under the guidance of the *Schiffer*, Philip Nagel. It was a dull morning, but it cleared up to a fine day. We rowed on until we came to a part of the river where it makes a great bend, and getting out opposite a small village on the right bank, we crossed the neck of land until we came opposite a little town, Neumagen, where our boat overtook us, and we re-embarked   About six o'clock in the evening we reached Berncastel, and here was the first great feature in the scenery; till then we both agreed that the scenery, although very pretty, had been over praised, but the town of Berncastel surmounted by its ruined castle inclined us to alter our opinion. We were told that we should sleep better at Trarbach than at Berncastel, and, therefore, taking a guide we crossed the land at that large bend, over a somewhat steep ascent, admiring the scenery very much, and reached the town sometime after it was dark. Our guide had been a conscript in the French army, and was in the fatal expedition to Russia. Our *Schiffer*, too, had been in the French army, was taken prisoner, and after some time entered the British service. Trarbach is a dirty old town with some pretty houses in the suburbs, overhung by Gräfinnburg, a ruin in a grand position. The bedroom was but indifferent, but I had a good clean bed. We had an excellent supper with the hostess, and her daughter and son-in-law, the Friedensrichter (a Herr Major who seemed to be a considerable person in Trarbach, a sensible, well-informed, well-bred man) together with the Friedensrichter's son, *Friedolin*, a nice boy of eleven years of age, with whom I had a great deal of talk before supper.

The servant volunteered to sleep in the boat and to come round the promontory with the *Schiffer*, and leaving Berncastel at twelve, they were round to Trarbach by five in the morning. It rained heavily all night, but was again clear in the morning, and we were off by six. The whole of yesterday was a succession of most exquisite scenery; the river makes another extraordinary bend, so that the neck of land may be crossed in a quarter of an hour, and it takes three hours to go round. This point is the finest in the whole course of the Moselle, and upon the promontory stand the ruins of the convent of Marienburg, at an elevation of about 400 feet. I have never seen anything finer than the view from this spot and we both agreed that there is nothing on the Rhine so·fine. We had some coffee at Alf, a small village at the mouth of the stream of that name, and then passed the beautiful points of Beilstein and Cocheim, at which place. we arrived about four ; and it is a most singular and beautiful spot. It was a fine afternoon, and we continued on our way to Carden, a small village opposite the little town of Treis, and where we found a most clean comfortable inn, like a very nice inn in the country of England. As I was desirous to arrive at Coblentz in time for the steamboat, we were up by four, and off by five. In less than an hour my companion left me, as he wanted to see an old castle of some repute, in the·interior.

Count St. Priest proved a most agreeable companion, accomplished, graceful, quite unaffected, and cheerful. His father was ambassador at, Constantinople under Louis XVI. when the Revolution broke out, and he retired to Russia with his family. My friend, the eldest son, entered into the Russian service, in which he must have risen to a high rank, for he was Governor of the province of Podolia, and afterwards of the Crimea, and he now possesses estates round Odessa. His daughter married the Prince Dolgorucky and he was on his way to Baden-Baden to see her. All this I learned yesterday morning from Francois, a most intelligent obliging fellow, who told me his own story, and he seems to have visited every part of Europe, even to Tulliallan, having been a travelling servant for a short time to Count Flahault, and he lived in the Murray's house in George Street. Count St. Priest has a great desire to see Edinburgh, and if he does, has promised to come and see me, and has taken my promise to visit him in Paris when next I go there.

I cannot describe to you how delighted I felt as I approached the ground I knew so well, as I came. nearer to Bonn. The first thing I noticed was Professor Bethman Holweg's new house at Rheineck, close by the old ruin, one of the grandest

situations on the Rhine. Then came Brohl, and then the
Drachenfels came in sight, and the long range of the Siebengebirge, with a thousand recollections. I found an old
acquaintance among the porters, and went to the Trierscher
Hof, and after taking some dinner went to the Harless's and
drank tea with them, and after an hour's talk over all the
good people of Bonn, I proceeded to the Windischman's, and
was most kindly received. God bless you all.

————

To his Wife.

Bonn, 12th September, 1835.

After finishing my letter to you yesterday morning, my very
dear wife, I sallied forth and looked into Marcus the bookseller's shop, and called on Nöggerath, who was in his usual
morning costume. He talked a little about the future meeting,
the arrangements of which he has the chief duty to look after.
I proceeded to Popplesdorf, the *allée* was in great beauty and
the view lovely. I saw Mrs. Bischof, and then visited the
worthy good old Goldfuss, who was as kind as possible, and
made many inquiries after you all. I returned to Bonn and
went to see Professor Arndt, whom I found at home, and
hearty and *loud* as usual; he is a good old man, with a fund
of knowledge of all sorts, and therefore very instructive conversation. I then went to Schlegel's, who received me most
kindly, and was very talkative and agreeable, he was in his
deshabille, with his black cap, the costume in which he looks
best; while we were talking, a note came from the Bishop of
London, offering to drink tea with him, an offer which he
accepted, then there was such a bustle, Heinrich and the
housekeeper were called up, the drawing-room was ordered to
be got instantly ready, *wax* lights to be set and tea to be
prepared. He said then to me " I must go and dress, for I
must not receive a Lord Bishop in such a costume," away he

went; I was near saying, "You will look much better as you
are." In ten minutes he came back, a gay dandy, with velvet
waistcoat and a gold chain, and calling out to Heinrich for his
gold snuff-box. We went down, and shortly after the Bishop
arrived. He had no episcopal costume. I left them in half-
an-hour. I saw the worthy Bleeks, the Sacks' live opposite,
but he is gone to Berlin, the Brandis are gone to Copenhagen.
I went to see Countess Beust and had a most agreeable hour's
conversation with her, I need not say the first part with all
whom I see, is answering most kind inquiries after you and
my dear girls, who live warmly in the recollections of the good
Bonners.

God bless you, dearest Anne,

Yours very affectionately,

LEONARD HORNER.

To his Wife.

Bonn, *19th September*, 1835.

MY DEAREST ANNE,—I sent off a letter on Thursday to tell
you of the arrival of our dear children (Lyells), that day
we went to dine at the general table of the Naturförscher.
Nöggerath did the honours, and on one side of him sat Von
Buch and on the other Elie de Beaumont. At one table we
had Dr. and Mrs. Buckland, Constant Prévost, Greenough,
Dr. Meyer from Bucharest, and two French gentlemen.
After dinner we left Charles with Elie de Beaumont, and
Mary and I took a walk in the Hof-Garten and to the Alten
Zoll. Yesterday was the first day of the meeting. Dr.
Meyer, of Bucharest read an interesting paper, giving an
account of the proceedings of the reigning Prince Ghika for
the improvement of Wallachia, with some geological
observations. I made the acquaintance of Professor Ritter
of Berlin, the geographer. The room was very full—places

had been previously assigned to the members, and I found
myself seated between Elie de Beaumont and Froricss, and
then Charles, Buckland, and Robert Brown. After the
meeting we went to dinner. It took place in the new ball
room of the Lese Gesellschaft. We got good places, and
there was a row of Berzelius, Charles, Mary, myself, Constant
Prévost, and Jussieu the botanist. In the evening we went
to Count Beust's, who has open house during the Versammlung,
and met there Alexander Brogniart and his son Adolphe.
I was very glad to be introduced to the old gentleman,
whose name has been very familiar to me these thirty years, as
the author of an admirable treatise on Mineralogy—but
rendered illustrious in the scientific world by his joint work
with Cuvier on the Mineralogy of the Environs of Paris,
wherein they first made known the assemblage of tertiary
rocks with the wonderful assemblage of the bones of extinct
animals. His son is Professor of Botany at the Jardin des
Plantes, and has devoted himself very much to fossil botany,
and with great success. This morning, Charles and I went
to Poppelsdorf, to the Geological Section, where we had a
strong muster. Von Buch in the chair, a paper by Dr.
Schmerling of Liège, upon the fossil bones found there in caves,
mixed up with human bones; another paper, by Von Hoff
of Gotha, on the curious footsteps of some animal, found on
the red sandstone of Hildburghausen, models of which,
Humboldt showed me at Paris. There was a good deal of
discussion, in which Buckland took a large share, talking
French fluently enough, but in utter defiance of all rules of
grammar and pronunciation. Yesterday there was a great
party to see the wonders of Cologne—but Charles and Mary
and I never thought of going, and I formed a party with some
of the French geologists, and we set out in two carriages.
Alexander Brogniart, Constant Prévost, Audion, (Brogniart's
son-in-law, and a most pleasing gentlemanlike man), and I in

the first, and in the second Jussieu the botanist, Robert Brown, Adolphe Brogniart, and Mr. Ampère. We went by Kessenich to Friesdorf, and stayed there two hours, examining the brown coal deposit with great minuteness; we then drove by Godesberg to Mehlem, where we crossed over to Königswinter, there we had some coffee in the Berliner Hof, looking over the Rhine, and as we were so employed, another boat arrived containing Bischof, Berzelius, and about a dozen more on an expedition to the Drachenfels, &c. Our party proceeded up the pretty valley from Konigswinter towards the Margareten Kreuz, which I daresay you remember well. I was the guide with my geological map in my hand, and I led them to the interesting points. All this occupied us till half-past three, when it was necessary for M. Prévost and me to depart for Godesberg for Dr. Mendelssohn's party. We crossed over from Königswinter to Rhings-dorf, and then walked across the fields to the Hotel. We found a large party sitting before the door drinking coffee, and waiting for us. It was a very fine day. Presently Mary came up to me and said, " There is a lady here wishes to see you." I turned round—it was Mrs. Mendelssohn, who had come down from Coblentz purposely for this dinner. She is looking very well, and the same engaging creature she always was. I had the happiness to sit next her at dinner. We had besides Charles and Mary, Dr. Mendelssohn, Mr. Bethman Holweg, Constant Prévost, Professor and Mrs. Dirichlet, of Berlin, Professor Plücher, and Professor Ritter of Berlin.

My kindest love to all my dearest children. Farewell my dearest Anne,

Your ever affectionate,

LEONARD HORNER.

P.S.—I was much pleased and interested by the French geologists, particularly old Brogniart, a little active old man of 67, and whom I have long considered with great respect as

an author. It is now half-past seven, and I have to call for
Charles at eight, to go to Poppelsdorf, where the Geological
Section is held.

24th September.—On Tuesday morning we had an excellent
discussion in the Geological Section on the great contested
theory of Von Buch, of Craters of Elevation. The discussion
was between Von Buch, and Elie de Beaumont, on one side,
Constant Prévost and Charles Lyell on the other, and was
carried on in French. Charles speaks with considerable
fluency. We arranged a party to go to the Rottenberg and I
was the guide. We set off from Buckland's lodgings at two
o'clock. In the first carriage were Von Buch, Buckland, Von
Oeynhausen, Elie de Beaumont, and myself, then came Mrs.
Buckland, Dr. Robertson, M. Römer, Constant Prévost,
Greenough, Lord Adair. We drove to Mehlem, and ascended
the Rotherberg, had long discussions on its mode of formation,
and then descended to Rolandseck, where we had coffee, and
got back to Bonn at nine o'clock. On Wednesday at
Popplesdorf, we had some interesting communications from
Dr. Abiot of Brunswick, a very accomplished young man, who
has travelled in Italy and Sicily, and brought many most
valuable facts away, concerning the eruptions of Vesuvius last
Autumn.

To his Wife.

Bonn, *24th September*, 1835.

MY DEAREST ANNE,—After the Geological Section, Charles
and I, accompanied by Greenough went to fetch Mary, to take
her to the General Meeting. In the library we found Mr.
Mendelssohn with her and we all went together. At the
dinner we had the Mendelssohns, Bethman Holweg, Constant
Prévost, Professor Ritter, Professor Dirichlet of Berlin and his
wife, next to whom I sat, and a very nice person she is, she is
sister of the famous composer, and cousin of Dr. Mendelssohn.

Brohl, 26th September.—I am sitting at an open window

looking upon Rhein-Brohl with the heaps of trap between my window and the Rhine. The object of this expedition is to visit the volcanic district of the Laacher See. The plan was formed by Von Buch. I went, yesterday, to breakfast with the Bethman Holwegs at Rheineck, within half a mile of Brohl; the situation is splendid, and the views up and down the river are among the best on the Rhine. Looking down from the Castle towards Bonn, we saw a long train of carriages, so Dr. Mendelssohn and I hastened down, and arrived at Brohl at the same time. There were between thirty and forty. A number of carriages had been brought from Andernach, char-â-bancs, &c., and some on foot, some on horseback. I was one of the pedestrians, preferring the opportunity of breaking a stone when anything interesting occurred. We went up the *Brohl Thal*, and when we arrived at Sonnenstein, where the mineral spring is, we found a collation laid out for us. After a pause of half an hour, we resumed our walk, and arrived about two o'clock at Kloster. Von Buch gave the whole party an excellent dinner in one of the large rooms of the Convent, and we sat down about seventy. I sat next Mrs. Buckland, she and Madame Gumprecht, a very pretty, accomplished and pleasing person, were the only ladies. I regretted the immense duration of the dinner—three hours—for it was exactly the best part of the day. We started at five, a grand cortège, for the Quarries of Niedermendig. I was in the carriage with Von Buch, Nöggerath, Professor Link and one or two more—it was a char-â-banc, with three *gigs* hung across. Having started so late we did not reach the quarries till it was nearly dark, that, however, was not of much consequence, as the principal thing was the underground. We descended into the mines, which were lighted up with a multitude of torches and had a most striking effect, the vast columnar structure, and the cross sections of the pillars gave the roof the appearance of a tesselated pavement. When we came to the surface of the earth again, it was quite dark and was raining heavily. We got into our open carriages the best way we could, and found the seats soaking, but there was no remedy. We had two and a half hours before us, ere we could reach Andernach. We had not gone far when the spring of the gig, on which I sat, broke, and I had a hard jolting ride the rest of the way. Half way we were stopped by a cry behind that one of the carriages had been upset, and shortly afterwards the horses galloped past, happily they were soon caught, and in about twenty minutes the cavalcade proceeded. We entered Andernach to the sound of music, and all the people were turned out to see us. The first thing was to get the best bed possible, and I contrived to do very well in that way; I dressed and in a quarter of an hour was very comfortable. Not so others, as I soon heard; the carriage that was over-turned contained Von Buch, Mrs. Buckland, and Madame Gumprecht, and poor Mrs. Buckland has suffered severely. I went to her room and found her in the hands of a surgeon. She had a severe cut in the cheek which had penetrated to the bone, and her forehead was swollen as large as an egg. I held her head while the surgeon sewed up the wound. I never saw a person behave with more beautiful composure and courage. She has passed a good night, and there is no fever, Von Buch was a little stunned and Madame Gumprecht escaped. Buckland and Greenough missed the carriage and had to walk in the rain from Laacher See to Andernach. It has rained all night and is raining heavily still. The party has separated for their several destinations. I am going with Dr. Mendelssohn to his father's place at Horcheim, near Coblentz, and I sleep at Coblentz to-night. My love to all my dear children.

Most affectionately yours,

LEONARD HORNER.

CHAPTER XIII.

1836—1837.

To his Daughter.

London, *23rd February,* 1836.

MY DEAREST FRANCES,—I was occupied the greater part of yesterday in the preparation of a paper for Mr. Le Marchant, Secretary of the Board of Trade, and dined at home. After tea Charles took me to the Geographical Society, where he left me to go to the Soirée at the College of Physicians. At the Geographical, there were some interesting communications on the proposed new expedition to explore farther the practicability of a communication between the Atlantic and the Pacific, by a north-west passage, following up the discoveries of Parry, Franklin, Ross, and Captain Back.

25th February—Yesterday Charles and I went to the Geological. Mary sent her carriage for us at ten, and we went home to dress, and got to Berkeley Square (Lady Lansdowne's) about eleven, and found a large number assembled. It was a brilliant sight, for almost all were in full dress. The greater number of the Ministers were there, I had a kind shake of the hand from the Premier,* and talked some time with the Chancellor of the Exchequer.† I saw a great many whom I knew. The Mintos, Lady Catherine Boileau, and her husband, Sir Charles and Lady Adam, the Locks, Admiral Fleming, Blakes, Hallam, Sydney Smith, Hollands, Rogers, and many others. I shook hands with Prince Czartoryski and Count Zamoyski. The Turkish Ambassador

* Lord Melbourne.
† Mr. Spring Rice, afterwards Lord Monteagle.

and his suite, very odd figures, were presented to the Landgravine of Hesse-Homburg, our Princess Elizabeth. I stood near the Duke of Wellington for some time, who is looking remarkably well.   We did not get home before two.

The great topic of conversation just now is the great triumph of good sense and good feeling in what passed in the House of Commons on Tuesday night, on the subject of the Orange Lodges; Lord John Russell's admirable speech, and his still more admirable measure.   He had the singular good fortune to reconcile the most opposite parties, and upon a subject too, upon which there was the greatest difference of opinions.   I spoke yesterday to an ultra Tory and an ultra Radical, and they both declared the speech and the measure to be admirable. Nothing has occurred like it in Parliament for years ; it will have the most beneficial effect on Ireland, and will give great strength to Lord Melbourne's government.   Do not fail to read the account of the debate, it will be in the *Scotsman* of Saturday.

Yours, my dearest Frances,

Very affectionately,

LEONARD HORNER.

To his Daughter.

London, 7th March, 1836.

My Dearest Katharine,— I was to have dined at the Hewlett's on Thursday, but ·have to be at the House of Commons that day, as the President of the Board of Trade* is to move for leave to bring in a Bill to amend the Factory Act; I suspect it will not pass quietly through, for there has been a meeting at Manchester of the operatives to petition·for a Ten Hours Labour Bill, and there is a deputation come up on

° C. E. Poulett Thomson.

the subject.   If the subject is gone into, I may be detained a long time in town. . . . . .

I have no public news to give you.   It seems to be the general impression that the Government is getting stronger every day.   There is to be a stout battle in the House of Commons to-night on the Irish Municipal Bill.   The chief subject of talk lately has been the controversy about the appointment of Dr. Hampden at Oxford, as Professor of Divinity.   The Tories there of the ultra cast, have discovered in a book of sermons he published four years ago, some opinions which they declare not to be orthodox, and have been moving heaven and earth to get his appointment annulled; but they will be obliged to retire discomfited from the field.

The appointments in consequence of the death of the Bishop of Durham are much talked of, especially the reduction in the revenues of the See.   Dr. Maltby, who is appointed Bishop of Durham, is to have £8,000 a year in place of £18,000 which his predecessor had.

> Ever, my dearest Katharine,
>   Your affectionate father,
>   LEONARD HORNER.

To his Daughter.

Athenæum, London, 17*th March*, 1836.

MY DEAREST LEONORA,—I hope soon to hear that you and Kate are pursuing your Italian studies with pleasure.   It is a charming language in itself, and abounding in authors of the greatest interest. . . . . Read in the "Biographie Universelle" lives that bear upon your Italian studies. While you are taking lessons, your thoughts should be mainly directed to Italian, but do not lay aside German; half-an-hour a day will "keep your hand in" as the phrase is.   I have often told you that the painter's maxim "*Nulla dies sine*

linea" is of universal application. I lament to say that the
chimpanzee died on Tuesday evening, and that without my
having seen him. Yesterday, after finishing my business at
the Board of Trade, I went to call on Sir Francis Chantrey.
. . . . He was just returned from Windsor, where he had
been spending two days with the King. In his studio I saw
his bust of Mrs. Somerville, which is quite admirable; I have
seen nothing which has pleased me more for a long time, it is
so very like, and so pleasing. It is to be placed in the
meeting-room of the Royal Society, a great honour at any
time, but especially during her life time. There is a very fine
statue of Bishop Heber. I had a long talk with Chantrey, when
chance led him to talk of his early life. He came to London in
1802 with a small independence left him by his father, and for
seven years he did not make a shilling by his profession. His
stock was growing low, and he felt the necessity of great exertion,
and so accomplished a work of some excellence ; it was a bust of
Horne Tooke. It was shown at the exhibition in Somerset
House, and such was the effect it produced, that before the
Exhibition closed, he had orders to the extent of £12,000,
that was in 1809. He said, "I was disposed to spend my
time in the luxurious ease of an artist, and if I had had the
misfortune to have had £100 a year more than I was possessed
of, you would never have heard of Francis Chantrey." In
the evening Charles and I went to the London University,
where we heard a most interesting lecture by Professor
Lindley* upon the evidence of the fossil plants in the coal
strata, shewing the existence, in that age of the world, of a
tropical climate in Northern Europe, an idea that has been
very generally adopted, but which he now declared, for the
first time, to be made out by the nature of the plants. There
was afterwards a meeting in the Museum, where I saw many
old acquaintances. I was very kindly welcomed by many. I

* Professor of Botany.

must now conclude, having to go to the Home Office, and many other places.

Give my best love to dearest Mamma, and all your sisters.

I am always, my dearest Leonora,

Your very affectionate father,

LEONARD HORNER.

[In June, 1836, Mr. Horner had an interview with Lord John Russell, and his district as Factory Inspector was changed from Scotland to Lancashire and north of England, and he determined to make his residence in London; accordingly the family left Edinburgh, which with the prospect of living near the Lyells and many old friends, gave them much pleasure, though they regretted leaving other dear friends behind them.]

Manchester, *July 12th*, 1836.

MY DEAREST MARY,—I cannot allow this day* to pass over without wishing you and Charles many happy returns of it, and expressing the happiness with which I always dwell upon the thought how eminently blessed you are in each other.

I have not yet heard of your Mamma's arrival in Edinburgh. I long to hear how she takes my proposal of throwing the whole burden of removal upon her shoulders. I left London on Friday morning, and travelling by Woburn, Northampton, and Leicester, reached Derby between 9 and 10 at night. I had written to Jedediah Strutt at Belper, to tell him of my intention of visiting their mills, and I found a messenger waiting my arrival, saying that his uncle Joseph expected me to sleep at his house, and that he hoped to see me next morning at breakfast at Belper. I found the old gentleman ready to receive me, and we had a pleasant talk of two hours before going to bed. Next morning I got up early to look at his collection of pictures and sculpture, which is extensive,

* Their wedding day.

and contains some very good things, and I got to Belper to breakfast. I spent the whole day there, was received with great kindness, and got some useful information about the management of their mills.

The Derby mail brought me here at half past three, and I went straight to Singleton Lodge, the house of George William Wood, who had kindly asked me to stay with him. He was Member for South Lancashire, and lost his seat at the last election, mainly on account of the active part he took about the Bill for the admission of Dissenters into the Universities. He knows Manchester thoroughly, and is highly esteemed. He was, therefore, best able to tell me the persons who are most likely to give a fair opinion on the various matters connected with the Act, upon which the right working of it depends. He wished me to take up my abode at his house, but as he lives three miles out of town, I told him it would not be convenient. Yesterday I had an invitation from the Borough Reeve (the Mayor of Manchester), to dine with him. It was to meet the Count Duchatel, who was Minister of Commerce lately in France, under the Duc de Broglie's Government. He is travelling with M. Duverger, a distinguished member of the Chamber of Deputies, I sat next the latter at dinner, and found him a clever, agreeable man. We had the *élite* of the mercantile aristocracy of Manchester. The Borough Reeve is a Scotchman, a Mr. Macvicar, who has been long in India as a merchant, and subsequently in London. His wife is a sister of W. Burn, the architect. Young Dr. Henry was there ; he came back six weeks ago from Berlin, where he has been working under Rose and Mitscherlich for eight months. He does not mean to practise, but to devote himself exclusively to chemical science.

I am, my dearest Mary's

Affectionate father,

LEONARD HORNER.

To his Daughter.

Liverpool, *July 17th,* 1836.

MY DEAREST FRANCES,—I have not written since last Monday, for I have had little leisure to do so. With regard to my duties I am meeting with great civility from the mill-owners, and from some I have met with great kindness. On Monday, I dined, as I mentioned I was going to do, with Mr. Macvicar the Borough Reeve. Besides the Comte Duchâtel and M. Duverger, to whom the dinner was given, there were about sixteen persons at table, some of whom are the most considerable people in point of wealth and influence in Manchester. On Friday I went to Pendlebury to dine with Mrs. Henry, we had a very nice party, Mrs. Henry, her son, Henry Romilly, Mr. Wm. Greg and myself. There is a large family of the Gregs of great wealth and great influence in this part of Lancashire; there are four brothers, all cotton spinners, and on a great scale; one has a mill near Wilmslow, another at Bury, another at Bollington, near Macclesfield, and the fourth at Lancaster. In all of them I hear the arrangements for the comfort and general welfare of their work prople are excellent. I had much valuable conversation with Mr. Wm. Greg. . . . . . Nothing can be more friendly and kind than Charles Henry, he has offered me his house to live in when I next come; I like him very much. They returned from Berlin only six weeks ago, where they have been living nine months, his great object being to work in the laboratories of Mitscherlich and Henry Rose, as scientific chemical research is his great occupation. His father is very rich, and he is an only son, so that he has no occasion to practise. Being desirous of seeing Mr. Melly, and his partner George Prévost, I wrote to the latter offering a visit, and got a very kind reply, so yesterday morning after breakfast I set out in one of the railway coaches. Mr. Wm. Greg told me he was going, and we arranged to go together. I found him at the station,

where he had secured places for us in the best carriage of the
train, the mail coach, and where we sat without any other,
as comfortably as in a gentleman's carriage. We were the
last of the long train of at least twelve, and saw the country
before us as we shot along, and were not at all annoyed with
dust or smoke. We were an hour and forty minutes, in going
the thirty-one miles; the motion is not only not unpleasant,
but is far superior to most stage-coaches. I had a most
agreeable companion, he is a man of great sense and information, and has travelled in Italy, Germany, Hungary, Sicily,
Greece, and Asia Minor. I have made an arrangement with
him, to go next Saturday to Bollington and some day next
week to Bury. I arrived in Liverpool about twelve, and went
straight to the counting house of Melly, Prévost & Co., where
I met with a most warm reception from both partners. I
walked about the town with George Prévost, and saw the
docks, which are truly magnificent. I was curious to see the
New York packets, of which I had heard so much, and we
went on board two; nothing can be better arranged for the
comfort of the passengers. I called on my old friend Wallace
Currie, who is the Mayor at present, and the first under the
new Municipal System. He was very glad to see me. He
was at the Town Hall, which is a very fine building, forming
one side of the Exchange, which is altogether very good—
superior, I think, to that of Amsterdam, and greatly so to
that of London, but not equal to the Bourse of Paris. It is,
however, on so different a plan from this latter, that there is
scarcely room for comparison. I am living in George Prévost's
house, a very capital one, in a handsome square. . . . .
We had at dinner, Mr. and Mrs. Melly, Henry Romilly, who
came from Manchester, and Henry Cockburn. Mrs. Melly is
sister of Mr. William Greg, and an agreeable gentle creature.
Of all the men whom I have seen since I came to Manchester,
there is no one with whom I am so much struck or pleased as

Henry Romilly, He is a delightful person, so gentle, so intelligent, so elevated in his whole bearing; he is a true son of Sir Samuel Romilly, more like his father in face than any of his brothers, though not so handsome or of so imposing a figure as Sir Samuel Romilly was. I return to Manchester to-morrow. God bless you, my dearest Frances.

Your affectionate father,

LEONARD HORNER.

To his Daughter.

Manchester, *7th August,* 1836.

MY DEAREST KATHARINE,—I had a meeting of mill-owners at Stockport on Friday, and dined that day in Manchester with Mr. Henry McConnell. He is the largest cotton spinner in Manchester, employing from fifteen to sixteen hundred people, paying not less then £1,500 every week for wages alone. He is a most excellent man, and I am glad to say he is going to adopt that which I am urging so strongly upon the mill-owners, the employment of the young children united with attendance at school. He is going to employ from two hundred to three hundred, and is to have a school upon his premises; he has advertised for a schoolmaster, and school-mistress, and says in his advertisement, that he wishes to adopt the system of the Edinburgh Sessional School. I had a talk with him, and his people, on Friday before dinner, and I have written to Mr. Wood,* asking him to select a master. His example will I have no doubt be followed. Yesterday morning I went to Macclesfield to breakfast, nineteen miles. I spent the whole day there, and returned here in the evening. Silk is the great manufacture of that place. I came home very well satisfied with my visit. There are a great many Catholics there; I was struck by seeing a

---

* Of the Edinburgh Sessional School.

placard in the street, with the following announcement :
" High Mass will be celebrated to-morrow at eleven o'clock,
when a selection of sacred music will be performed from the
works of the most eminent composers, among which Mozart's
celebrated mass, No. 12, and Zingarelli's Laudate, with full
orchestral accompaniments ; for the benefit of the Catholic
Schools."

I hope, dearest Kate, that you, Nora and Joanna will take
great care of yourselves in your Mamma's absence.   Love and
kisses to your sisters from

Your affectionate father,
LEONARD HORNER.

To his Daughter.

Halifax, August 14th, 1836.

MY DEAREST LEONORA,—I daresay you were all much shocked
and distressed on hearing of the sad death of Mrs. B. Praed.
It is very melancholy to see a person with every prospect of a
long and happy life before her, so suddenly cut off; a daily
lesson almost, if we would attend to it, how necessary it is to
be prepared for that awful change, in the midst of the highest
health.   On Thursday I went to Rochdale, where I passed a
day.   Near Rochdale I saw the seat of your great-great-grand-
uncle, Sir William Horton, Chadderton Hall, but I only saw
the woods.   Rochdale is a pretty country town, in a pretty
situation.   From thence I came on to this place, by Todmorden,
through a beautiful valley, deformed however by numerous
factories, which are not like Temples dedicated to water
nymphs by crystal streams, in all the grace and beauty of
Grecian architecture, but huge ugly masses of building
dedicated to water wheels, with the addition of a tall chimney
vomiting forth smoke, for the stream not being powerful
enough, they take the liberty to boil a little of the crystal
fount, and by making it force its way through a steam engine

they make the stream do their work in one way or another. These mills destroy the romance of the valley, but they bring a great deal of substantial comfort to the people. I never went through so thickly peopled a country, and there is no appearance of poverty, it is not the swarms of creatures you see in Ireland, so wretchedly off both for food and raiment, that it would have been far better if they had never been brought into existence; but a well-fed, well-clothed people; neither rags nor dirt; no mud dwellings without a chimney, but substantial stone and brick houses, of two stories, dry, clean, and well furnished. In the twenty-seven miles from Manchester to Halifax, I only saw one beggar, and he a very old man. At Rochdale my Superintendent, Mr. Trimmer, met me. It is a great place for the manufacture of flannel. I found a manufacturer there disobeying the law so seriously, that I had to order him to be prosecuted.

I was very much amused yesterday in looking over a list of factory children to find one called Xantippe, I thought that name had been dropped since the wife of Socrates made it so famous. I learned that she was the daughter of a weaver who is passionately fond of ancient history, and they shewed me in another list, the names of two other of his daughters, Diaphantes, and Pandora; but only think what a word was added to each, a word which the poor weaver could neither change nor modify: Barraclough—Pandora Barraclough!

I am, my dear Leonora,

Your affectionate father,

LEONARD HORNER.

————

To his Wife.

Manchester, *20th June*, 1837.

So the poor King is dead. I was indeed very much shocked to hear of dear Ellen Hallam's death, and most sincerely do I sympathise with the excellent and afflicted parents. I often think when such events occur, how I should be able to bear such a privation. We have indeed great reason to feel deeply grateful to God for his goodness in sparing us hitherto from so dreadful a blow, and we must live in humble hope, that we shall be permitted to precede all our dear children in the transition to another state of things. I would write to Hallam, but it is a very painful thing to do—and what can be said—and to such a mind as his, but I should like very much for him to know that I sorrow much at his bereavement.

I hope the Court mourning will be very short, for it does a vast deal of harm to tradesmen, and is a vain ceremony, because while one side of our face is weeping over the departed King, the other must wear a smile of congratulation on the accession of the young Queen. Our friend, Dr. Clark, will rise into much higher consideration now, for the Queen's physician-in-ordinary is sure to be advanced in his practice, especially when he is so deserving of it as he is; I expect he will be made a Baronet ere long. The change will greatly strengthen the present Government I should think, and if they are in *Court* favour, which they never have been yet, it will bring round a vast number of adherents; their majority in the House of Lords will soon increase.

I have been this evening to the meeting of the Statistical Society, of which I am a member, and met several of my Manchester friends, among the rest Henry Romilly, who has just come from Dublin, where he has been visiting the Kennedys. They have got a very comfortable house near Dublin; he is very well, but she is only better, having had a very severe illness.

* * *

To his Daughter.

Manchester, 25*th June,* 1837.

MY DEAREST LEONORA,—I have just returned from Barlow Hall, Mr. Shakspear Phillips's, where I went yesterday to

dinner, about five miles from Manchester. Till within the
last thirty years, the place had been in possession of an old
Catholic family of the name of Barlow for centuries, even
from the conquest it is said, but they are now extinct. Upon
a pane of stained glass in a very old-looking oriel window,
there is the date 1574. Mrs. Phillips is very agreeable, and
is a great favourite of mine. I went with them to Charlton
Church this morning, the curate had got his lesson well, for
he brought in Victoria at all the proper places, and never said
his for *her*. I leave to-morrow for Bury, where Mr. Phillips
has given me a letter of introduction to Mr. Hornby, the
Rector; he is brother-in-law to Lord Derby. It is one of
the best livings in England.

On Friday I went to Hyde, a large and densely peopled
village, about seven miles from Manchester, and I visited a
very large mill belonging to a Mr. Horsfield, a man nearly
seventy years of age, who is said to be worth at least £300,000,
and can hardly write his own name. I was curious to hear
from his own lips something of his history, and so entered
into conversation with him. He took me to his house hard
by, as he was going to dine, it being *twelve* o'clock. He had
a piece of cold beef and potatoes, no wine; he keeps one
woman-servant, and his daughter, whom I saw, was not much
in appearance above the maid. He told me that at eighteen
years of age he had not five shillings in the world, beyond his
weekly wages; that out of his wages of fifteen shillings, he
saved £28, bought a spinning jenny, and made £30 the first
year. In 1831, he made £24,000 of profit; he employs about
1,200 people. His is not a solitary case; there are many not
very unlike him in this part of the country.

The weather is beautiful, and the country is in the greatest
beauty. I walked through a grass field to-day, the fragrance
of which was quite delightful.

I suppose you are all much interested in the proceedings

of the young Queen.   I see by the newspaper that she has
sent for Lady Lansdowne and Lady Tavistock to be near her
person.   She has made a good choice independently of all
political considerations.   She will have a very amiable and
sensible person about her in Lady Lansdowne.   God bless
you, dearest Nora,

Your affectionate father,

LEONARD HORNER.

To his Wife.

Haslingden, *June 29th*, 1837.

Soon after my arrival in Bury I went to visit a mill, on my
return to the inn in an hour, I found Mr. Hornby's card and
a most kind note asking me to take up my quarters at his
house during my stay in Bury.   I met with a most kind
reception. I found him a most accomplished, modest, benevolent
man, in religion gentle and tolerant, in politics a good rational
Whig of the old school.   He is beloved in his parish, and it
is of enormous extent, containing a population of 50,000 souls.
She is also very agreeable, gentle, kind and sensible, with a
great deal of good lively conversation.   They have several
grown-up children, but the only one at home, was their only
daughter.   They have an excellent house quite in the town
of Bury, but in a large garden, and completely screened from
all disagreeable sights.   I remained there till this morning,
and they kindly said at parting that the more frequently I
came, the happier they would be to see me.   I reached this
place, about eight miles distant from Bury, about five, having
been visiting mills all the way.   It is a most lovely valley, at
the bottom of which is the Irwell, the ground in gentle swells
and well wooded, with the most rich verdure.   The most considerable establishment I visited is that of William Grove and
Brothers; three Scotchmen from Morayshire, who came to

Manchester forty years ago with nothing, and who are all
now immensely rich, but they are as liberal as they are rich.
They have about 3,000 people dependent upon their works.
Among other magnificent things they have done, they have
built, at their sole expense, a large and beautiful church with
a tower and bells, and strange to say, it is a Presbyterian
church, and they have a full congregation. The weather is
splendid, and although the sun is powerful, there has always
been a most refreshing air and everything in the fields wears
the aspect of an abundant harvest.

God bless you all.

To his Daughter.

Bowness, July 8th, 1837.

My Dearest Frances,—I am sitting at an open window
with Windermere Lake in all its beauty spread out before me,
in splendid sunshine with a delicious balmy air. As official
duties brought me within five miles of this delightful spot, I
could not resist the temptation of passing the half of this day,
and a quiet Sunday to-morrow, in this beautiful temple of
nature ; some compensation for the many hours I have lately
passed in the hot, close factories.   Bowness is about the
centre of the Lake, and nine miles from Kendal.   I am at the
Crown inn, which stands in a garden upon an eminence commanding the whole extent of the Lake.   Immediately below
me is the village, consisting of a number of neat, white, blue-
slated houses, scattered among trees, the church a neat plain
building, with a low tower, and three yews in the churchyard,
and behind a sunny round meadow, covered with fresh-made
hay ; this forms the fore-ground to the landscape : beyond, the
lake, smooth as a mirror, with several green and wooded
islands, and here and there a boat, of which I see the feathering
oars, but hear not a sound.   A wooded low range of hills
jutting out into the lake in many promontories, forms the

nearer background, and farther off a succession of mountains in hot, hazy obscurity terminates the prospect. In this spot I shall sit during the heat of the day, feasting my eyes with the positive beauties so amply spread before me, and my ears with the negative charms of the most delicious stillness.

If I had your dear Mamma and all of you here, to participate in my enjoyment, how vastly it would be increased. In the cool of the evening I mean to have a row on the lake. I had a delightful treat last night at Kendal, three budgets from home, and two of them containing letters from the dear travellers.* I hope they are at this moment enjoying themselves in the neighbourhood of Christiania. On Thursday I went to Caton, where Mr. John Greg resides. He was good enough to send his drosky for me. Caton is a village about four miles from Lancaster, in the vale of the river Lune. It is a most beautiful valley, and the view from what is called the brook of Lune is splendid, it is terminated by the lofty mountain of Ingleborough. Mr. Greg's house is charmingly situated. I arrived about two o'clock, and found Mrs. Greg sitting with her children at their dinner.

To his Daughter, Mrs. Lyell.

London, *August 20th*, 1837.

MY DEAREST MARY,—Last night when we returned from Babbage's, we found your most agreeable letter ; it is very pleasant to see how much the visit to Copenhagen has answered the object of Charles in going there, and how agreeably you have both spent your time. You will be glad to visit Copenhagen again, and I observe that with your usual energy, you are preparing the way by studying Danish. When I came to town I expected that after a week or ten days' occupation with my colleagues, I should have had a month or five weeks'

*Mr. and Mrs. Charles Lyell.

holiday, but scarcely a day has passed without my having had a good deal to do. I shall be a fortnight longer in town, and then shall start for Liverpool, during the Association meeting; and am to be at George Prévost's. I shall not return home till towards the end of November. We have had a good deal of pleasant society, and have seen more of Babbage than anyone. We have seen Darwin several times; not so often as we could have wished, but he is working so hard that he does not go out. He had not when I saw him received any part of his MSS. in type. During the week that Dr. Forchhammer was in town lately, we saw him four or five times, and were much pleased with him. I have not seen anyone for a long time with a greater store of accurate knowledge.

We have been two expeditions to Sir James South's at Kensington, we saw several double stars and Saturn in broad daylight, and the latter in great splendour after sun-set. We saw three of the most remarkable nebulæ. He was most kind and obliging.

On Tuesday I took Frances, Nora and Joanna, to Loddiges, where Stokes met us, and they had an excellent botanical lesson.

Mr. Falck was lately in London, and I passed an hour most agreeably with him. He would have dined with us, but I did not see him till the day before he set out on his return, He came to consult Alexander the oculist. He spoke a good deal to me about a visit M. Cousin made to Holland last year, to examine into the state of public instruction there, and excited my curiosity so much, that the idea occurred to me of translating his report as Mrs. Austin did that on Prussia, as this subject is now so much before the English public, and *is* likely to be brought before Parliament next spring. I therefore wrote to Count Flahault, to apply to Cousin for a copy of his report, and I received it yesterday with a very civil note, saying that he had sent the first copy he had got of his work from the bookseller, and that it will not be published for a fortnight. What I have read of it is very interesting, and shews the liberal and enlightened views of the Dutch Government in all that relates to education and the perfect toleration on religious matters for which that country has been so long distinguished. But the size of the volume alarms me, four hundred and sixty octavo pages. One main inducement is, that in the shape of a preface, I could give some of my own views in reference to the establishment of schools for the working classes.

We go to the Somerville's to-morrow. Miss Rushworth, who was at our house on Wednesday, gave us a comfortable account of the Hallam's, with whom one of her sisters is staying at Sevenoaks. Give my kindest love to Charles.

Believe me, my dearest Mary,

Your affectionate father,

LEONARD HORNER.

———

From Mr. Hallam.

Sevenoaks, *8th September*, 1837.

MY DEAR HORNER,—I have abandoned all thoughts of going to Liverpool. Whatever gratification I might feel in it, the long journey, the continued bustle, and above all, my reluctance to leave home during Harry's holidays, deter me from the attempt. Another year I may feel more equal to the meeting. You have my best wishes that it may now pass over agreeably to all my friends who assemble there.

We are all in perfectly good health. My wife has borne the great affliction with admirable composure, and resignation; but the world can never be to us what it was before these repeated severances of those we best loved in it, and it is doubtless better that this should be the case. We shall continue at least for the present to live in London; but mingling less, probably, in general society than you have seen us do for some years past. For myself, I shall more and more devote myself to those intellectual pursuits which withdraw me in some degree from the recollection of domestic sorrows, and to the conversation of those who partake my relish for them.

Believe me, my dear Horner,
Very truly yours,
HENRY HALLAM.

To his Daughter.

Liverpool, 10th September. 1837.

MY DEAREST FRANCES,—Galton, Emma and I started in the train from Birmingham, and arrived here at four o'clock, having travelled in four hours and a half, ninety-seven miles. We went straight to the Town Hall to get information about the proceedings, and procure our tickets. I learned that the general committee had appointed me one of the vice-presidents of the Geological Section, with Lord Cole and Mr. De la Beche for my colleagues.

At ten we went again to the Town Hall, where the Mayor's Soirée took place. There is a large suite of beautiful rooms, with a grand staircase, on one flight of which stands Chantrey's marble statue of Canning, who was for some time member for the town. I found Sedgwick, Murchison, Greenough, Sir Philip Egerton, Lord Cole, De la Beche, Roget, Whewell, Lord Burlington, Lord Northampton, and many others.

Monday.—Yesterday I went to church with his Worship; Dr. Roget and I walked arm in arm in the procession. There was most beautiful music. After the service I went to the Athenæum, where there were a great many of the wise men, and new ones dropping in continually, Professor Henslow, Griffith from Dublin, Peacock from Cambridge, &c., &c. On arriving at Prévost's to dinner, we found De la Rive, who came yesterday, and we had a most agreeable afternoon. De la Rive is a man of great and varied knowledge, and whose conversation is excellent.

Sir Philip Egerton has asked me to go to his house after the meeting, and I have accepted.

My love to dearest Mamma and all,
Your affectionate father,
LEONARD HORNER.

To his Daughter.

Liverpool, 14th September, 1837.

MY DEAREST SUSAN,—Tuesday morning was spent in the Geological Section, and in the course of it I read Charles (Lyell's) paper on the neighbourhood of Christiania. At three, Lord Cole and Professor Lindley and I walked to the Docks. I got home in time to dress, and reach the house of Mr. Heywood, two miles out of Liverpool, by six o'clock. There was a large party of the wise men, and I had the good fortune to sit between Professor Moll and another most agreeable man, whom you may remember at Edinburgh, the Rev. Dr. Lloyd of Trinity College, Dublin. At nine Mr. Lloyd and I went together to the Town Hall, where there was an assembly, and the large and beautiful suite of rooms were filled to excess. Yesterday, after finishing my factory business, I went to the Committee of the Geological Section for an hour, and then to the section of the Statistical Department, where I heard a paper read by Mr. William Greg on the state of the working classes in and around Manchester.

I went back to the Geological Section, when I heard the account by Sedgwick of the calamity at Workington. There was there a coal mine, which had been worked for many years far under the bed of the sea, so that there were vast under-

ground excavations and miles of cracks through it, with
rail-roads. The overseer most rashly and most culpably
carried the workings far too near the bed of the sea, and
although he had such warnings of the oozing in of salt water,
that many of the men remonstrated, he persevered, and even
turned off some old miners who refused to go into the works.
On a sudden the sea broke in, and in three hours the whole
works were filled to the pit's mouth, and property yielding
above ten thousand pounds a year, was thus irrecoverably
lost. Fortunately it happened at a time when the greater
number of the workmen were above ground, but twenty-nine
lives were lost. Sedgwick's description was most graphic, and
particularly of the escape of a miner of the name of Brenner,
who with great intrepidity and at the greatest peril to himself,
saved the lives of two others. The story was quite affecting,
and there were abundant proofs that the hearts of Geologists
are not made of granite, for both Sedgwick and several others
were moved to tears. The effect upoon the audience was
quite striking; and upon the instant, Murchison proposed a
subscription for poor Brenner, hats were sent round the room
in all directions, and thirty-four pounds were collected in
shillings and sixpences, which Sedgwick carried off in triumph,
and he was to write last night to Brenner under a frank from
Lord Burlington. After the meeting I had a long talk of an
hour and a half with Mr. Cheetham and Mr. Ashton, two of
the most intelligent and benevolent cotton spinners near
Manchester, upon the subject of the education of the factory
children. I then went to dine *en famille* with George Prévost,
and he, his wife, John Prévost and I, went in the evening to
the Amphitheatre, where a very interesting lecture was given
by Snow-Harris of Plymouth, upon his mode of protecting
ships from lightning, illustrated with many beautiful
experiments. The theatre was crowded to the roof, and the
stage was occupied by the Life Members of the Association.

I had a good seat in front, next to Mr. Richardson,* who went
with Franklin on the Polar expedition. I came away with
Mr. Jones Loyd the banker, and Dr. Arnott, who brought me
home.

I am, my dearest Susan's .

Affectionate father,

LEONARD HORNER.

——

To his Daughter.

Oulton Park, 19*th September*, 1837.

MY DEAREST KATHARINE,—Yesterday morning after de-
spatching some factory business, I went to the railway
station, where by appointment, I met Whewell, De la Beche
and Clarke. We arrived at Hartford, about five miles from
this place, where we found a barouche sent by Sir Philip.
It was unfortunately a rainy day. We drove over a somewhat flat and not interesting country, with a considerable
extent of heath, for about five miles, when all at once we
emerged from this wild heath into a beautiful park with the
most luxuriant turf and noble trees; a most striking contrast,
and showing the great age of the park; for the soil that
produced such trees and such turf must have been the
accumulation of decayed matter for a long period of years.
In half a mile we reached the house, which is a very large
square building, partly brick and partly stone, surrounded by
a broad terrace; it was built in the reign of Queen Anne.
The park view from the house on all sides is good, and I now
see a large herd of deer from my window. We came in just
at luncheon time; we were met by Lord Cole at the door, in
a rough jacket and trousers, and a white rough felt cap, for
hat it cannot be called, which he said he bought for eightpence;
and, by his side, our old friend *Jack*,† who is in great vigour.

* Afterwards Sir John Richardson.

† Lord Cole's large dog.

We met with a most kind reception from the host and
hostess, and found besides them, Miss Leigh, her sister, and
a Mr. Dupré and his two daughters from Buckinghamshire.
There is a most splendid entrance hall, and the rooms in
general are very handsome, particularly the dining-room, and
there are many good pictures. In the dining-room there is a
portrait of Lord Strafford by Vandyke, and a most singular
picture, Quin the Actor in the character of Cato, in a court
dress with a full bottomed wig, and this celebrated gourmand
and epicure, resting his hand on Plato's work on the
Immortality of the Soul, in the form of a thick folio with
silver clasps. After luncheon and a survey of the pictures,
we went to Sir Philip's Geological Museum, where we stayed
an hour, looking at some of his most remarkable things. He is
particularly rich in fossil fishes and fossil bones, and has
very many most interesting specimens. He has a very
accurate knowledge of his subject, and devotes a great deal of
his time to it. He will make an excellent President of the
Geological Society. Sir Philip, Lord Cole, Whewell, De la
Beche, Clarke, and I sallied forth to the park afterwards, for
it had cleared up to a fine afternoon, and bent our way
towards a lake, where lines had been set to catch some pike.
We got into a punt, and while Whewell rowed, Sir Philip and
Lord Cole drew in the lines, De la Beche received the fish,
and put them out of their pain, and Major Clarke and I sat at
the stern as lookers on, with Jack between my legs. We
found five pikes, two of them very large. We then walked
about the park, and Sir Philip showed us a lime tree, with
feathering branches to the ground, which Clarke and Whewell
calculated might shelter under its boughs a regiment of
one thousand two hundred men. It puts one in mind of
the great chestnut tree in Etna, called the Cento Cavalli.
When we got back to the house, we found Lord Northampton
arrived with his son Lord Compton, and Lady Mary Anne

Compton, his daughter, all of whom had been at the Liverpool meeting. Soon afterwards we sat down to dinner, all the persons I have named, with the addition of Mr. Cornwall Leigh, Lady Egerton's brother, and Capt. Herbert Taylor, her brother-in-law. In the drawing-room we had some good music from Lady Egerton ; but particularly from Lord Compton, who, with his sister, sang some things out of the Puritani. Lord Compton's voice is magnificent, I never heard anything like it except Lablache ; and Clarke, who is a musical man, said the same thing. Lord Northampton is a most agreeable person, and his son is a very unaffected and sensible young man. I am very much pleased with Lady Egerton, she is very pretty, and has most pleasing manners. He is a capital man in every sense, and would make an excellent Whig, if he were not a Tory. To-day we are to have an expedition to the salt mines near Northwich, where I was thirty-one years ago. God bless you all.

My dearest Kate's most affectionate father,

LEONARD HORNER.

———

To his Daughter.

Oulton Park, *20th September*, 1837.

MY DEAREST LEONORA,—After finishing my letter to Kate yesterday morning, I went to breakfast, and found all the party, very nearly, already assembled in the dining-room, it was a merry pleasant meal, and at the conclusion of it, Lord Cole amused us with some exploits of Jack. The children came in, and the eldest, a boy of four years old, Philip, is an uncommonly fine child. We then took a walk in Lady Egerton's flower garden, which is uncommonly pretty, being a mixture of shrubbery, forest and flower bed, which has an excellent effect, and in some parts a view of the fine facade of the house comes in. How much you and Frances would have enjoyed

the greenhouse, for there are a great number of curious plants,
such as we saw at Loddiges. There is one large bed entirely
composed of fuchsia, which is just now in the greatest beauty,
and there is a large standard magnolia on the lawn. About
half past eleven, we set out in four carriages, with Sir Philip,
Lord Compton, and Mr. Leigh on horseback, for Northwich,
to visit the salt mines. I was in the britska with Lord Cole,
Whewell, and De la Beche. The proprietor of the mine was
ready to receive us, and Sedgwick and Murchison had arrived
from Liverpool to give us the meeting. We descended the pit,
which is three hundred feet deep, in the bucket by threes, and
when we got down, we found the mine lighted up in every
part, by two thousand three hundred candles; the exca-
vated space is equal to twenty-four acres: and huge square
pillars of salt are left at distances to support the roof. Lord
Cole had brought with him some Bengal lights, and when they
were set fire to, the effect was much increased The miners
made some blasts, and a small cannon was fired, and the sounds
reverberated through the cavernous spaces like low thunder.
The place is perfectly dry, and of a moderate temperature. In
some places there is a strong reflection from the salt, but if
the accounts of the salt mines of Wielickska be correct, the
brilliant effect there must be greater. The salt in Cheshire is
mixed with a good deal of clay, which renders it less brilliant.
We ascended in the same way, having been down above an hour,
and then saw the process of salt making; the rock salt is
dissolved in water, and the earthy admixtures are allowed to
subside, and then the brine is evaporated, and while it is
going on, the fluid is disturbed, which prevents the formation
of large crystals, which would take place if the brine were left
still. The salt is then taken up by large iron spoons with
holes to let it drain, and put into square boxes, and these are
taken off in a short time, leaving large quadrilateral pyramids
of salt, truncated at the end, which are taken to a stove-house,

where they are exposed to a heat of about one hundred and thirty degrees Fahr. Having finished our examination, we set out on our return, Sedgwick and Murchison being added to our britska, and after a merry drive we reached home about half-past five. I sat at dinner next to Lady Egerton's brother, Mr. Cornwall Leigh of High Leigh, and found him a very agreeable man. The evening passed merrily, and as we were all taking our bedroom candles in the great hall, Lord Compton and Major Clarke gave us some songs which sounded well.

Thursday. Our party at dinner yesterday was reduced to fourteen, as Whewell, De la Beche, &c., were gone, and we had only got Greenough in addition, who came up from Liverpool yesterday. A discussion arose whether the figure of a Scotch reel is ∞ or ∞∞, led to an experimental enquiry, when Sedgwick proved his assertion that the latter is the figure, and that led to other dancing among the younger part, and they showed themselves well-skilled in the waltz.

This morning Sir Philip and Murchison are gone shooting. Sedgwick, Greenough and Clarke are gone geologising, and I terminate my very pleasant visit at one, and I expect to dine in Manchester.

———

Your affectionate father,
LEONARD HORNER.

To his Daughter.

Keswick, *September* 26th, 1837.

Well, my dearest Joanna, I have been to-day to the top of Skiddaw, and it is a thing to remember for the rest of one's life.

Yesterday morning after breakfast, I started at eight o'clock by the Whitehaven coach for this place. The vale in which Kendal lies is very pretty, and the road continues over a succession of low hills until we come to the borders of Windermere. At a high point of the road, the lake came at once into view,

and I again saw the magnificent prospect which delighted me
so much in July; and about a mile off I descried my nice
country inn, the "Crown" at Bowness. About two miles
from Ambleside we stopped at another beautifully situated
country inn, Low-wood, which is at the upper end of the lake,
and on its borders. There is here a very fine view up Langdale, behind the village of Brathy. In the distance I saw the
lofty mountain called Coniston Old Man, 2,577 feet high, and
nearer, the Langdale Pikes. This part of the lake is undoubt-
edly very fine, but I prefer, upon the whole, the views at and
near Bowness, particularly that from Brunt Fell, immediately
behind the Crown Inn. The scene changes quite at Ambleside,
which is a mile from the lake, in an amphitheatre of green
hills, rocks, and trees. We soon came to Rydal Water, near
which we passed the house of the poet Wordsworth. Rydal
lake is small, but there was an exquisite picture worthy of the
pencil of Cuyp, and the fore-ground of cattle standing tran-
quilly in the water at the edge of the lake, was not wanting.
A mile farther we came to Grasmere, which has quite peaceful
beauties of its own, and then we began a steep ascent, called
Dunmail Raise, a pass 750 feet above the sea, and which
separates Derwentwater and Thirlmere from Grasmere.
When we got to the summit of the road, Skiddaw's peaks
appeared in view, and the great Helvellyn was on our right.
A succession of striking points of view occurs all the way to
Keswick, and one of the most remarkable is that of the Vale
of Keswick. I got there at one, and I immediately called on
Mr. Edmondson, the surgeon who grants the certificates of
age to the children in the factories. I came here to inspect,
in order to arrange the order of my visits with him. He said
that he was just going to visit a patient at the head of Derwentwater, the usual name of Keswick Lake, which would
chime in very well with my route, and proposed to accompany
me; so I hired a car, and we set out together. But first I

must tell you that as I entered the town I noticed the church
which Mr. John Marshall, who died last year, was building
on his estate at the time of his death. It is a most beautiful
design, and the exterior is nearly finished. It is of polished
stone, which is brought from the neighbourhood of Penrith,
eighteen miles off. He was not at the whole expense, for the
chief part of the money was supplied by the Parliamentary
grant for building churches, but I suppose Mr. Marshall's
share could not be less than £3,000, from the elegance of the
building. The site is admirably chosen, for it is seen on every
side. Mr. Marshall, five years ago, bought from Greenwich
Hospital a large part of the Derwentwater estate, for which
he gave £60,000. This was the estate of the Earl of Derwentwater which was forfeited on his joining in the rebellion of
1715, and it was given to Greenwich Hospital. Mr. Marshall's
death was a great misfortune to this part of the country. I
drove with Mr. Edmondson quite round the lake; we first
passed the house of Mr. Pocklington under Wallow Crag, and
then the celebrated Waterfall of Lowdore, which I got out of
the car to see. It is a lofty fall, not in one sheet, but over rocks
and in a deep wooden chasm, but there was very little water,
it must be very grand in floods. We next passed the little
village of Grange, so called because it was a granary for the
supply of the monks of Furness Abbey. We now came to Mr.
Edmondson's patient, at a small farm house. It was a poor
man of 80 years of age, who had been brought a few days ago
from Egremont, to the house of a relative, in a state of furious
madness. I went upstairs with the doctor and saw the poor
maniac in bed, and it was a sorry sight; I stood by while the
doctor bled him, and put on a blister, but he will probably die
in a few days. We then drove along the lake to Stair, where
I had to visit a mill, and thence to the beautiful little village
of Braithwaite, where I had another mill. We had now just
time to get home in the dusk. It had been a magnificent

day, and I lamented that I was not on Skiddaw, whom I saw
before me in all his majesty, clear to the summit. I hoped
that the next day would be as good, and engaged Will
Grave, the guide, to go with me. I looked out this
morning soon after six with some anxiety, and it was
a clear frosty morning. At half past seven Will came, and
said that it was not a very good day for Skiddaw, but that it
would do ; adding, however, that he could not go, but that
his brother would, and that he was a better guide than himself.
At eight o'clock we started, John Grave and I. It is six miles
to the top, but I walked quietly, the day grew better and
better, and new and splendid scenes opened up as we ascended.
When about 500 feet high, there was a very fine view of the
lake, bright as a mirror, with the mountains and neighbouring
rocks clearly reflected in it. In three hours we reached the
highest point, and I was amply rewarded. It was a glorious
day, a bright sun, and the heat kept down by a refreshing
breeze. The guide said that there had not been a better day
for Skiddaw the whole year, and that not once in a hundred
times is it so good. I will not attempt to describe the view,
the mountains below me were like a vast ocean in a storm,
the waves tossed up in every form and direction. It was so
clear that I saw Criffel mountain in Dumfrieshire, Ingleborough and Yorkshire near Settle, and the Isle of Man, and
if you look at the map you will see the extent of the radius
of view. I quitted the summit with great reluctance. It is
3,022 feet above the sea, and 2,800 above Keswick. We
stayed about an hour on the top, and I brought away a
specimen of the slate with a beautiful lichen upon it from
the highest point. We descended in two hours, and I was
surprised to find myself so little fatigued. As soon as I got
home, I dressed, ordered a car, and went out to visit three
mills, four miles off, at Thornthwaite, Millbeak, and Apple-
thwaite. The road between the two last mentioned places

runs along the foot of Skiddaw, at an elevation of about 100 feet, and it commands by far the finest view I have yet seen, either here or at Windermere. A grand circle of mountains the full expanse of Derwentwater and the town and vale of Keswick, forming the foreground. A fine setting sun shewed it off to the highest perfection. I got home by six, much delighted with my day's work.

> I am, my dearest Joanna's
> > Affectionate father,
> > > LEONARD HORNER.

To his Wife.

Carlisle, 5th October, 1837,

I have stayed longer here than I intended. I was at home all Monday writing, for I drew up my quarterly report to Lord John Russell. After a walk I went to dine at the Deanery, and found them all alone. I passed a very sociable afternoon, Miss Hodgson sang some ballads very prettily and with good expression. . They proposed to take me a drive next day to see the great viaduct of the new railway over the Eden at an elevation of ninety feet, and Corby Castle, Mr. Howard's place. I readily agreed, because I had to visit a mill in the same direction. On Tuesday, therefore, after having been engaged in visiting mills in Carlisle till twelve o'clock, I went to the Deanery, and the Dean and Miss Hodgson and I set out, in their drosky, drawn by two ponies, the Dean driving. We stopped at Warwick Bridge, five miles off, where my mill was situated. We stopped at the house of Mr. Dixon, the proprietor of the mill, and Mrs. Dixon, a very nice lady-like person, who had been apprised of our coming by her husband whom I had met at Carlisle in the morning, had a nice luncheon ready for us. We then *all* proceeded to the mill, about a quarter of a mile off. I did not expect to be detained above a quarter of an hour by my examination, but finding some irregularity about the employment of two children, which I could not leave uninvestigated, and having to send for their parents, we were kept nearly an hour and a half, so that we were obliged to give up both the viaduct and Corby Castle.   There was a party at dinner at the Deanery ; we had some fine Italian music in the evening, by Mrs. Lake, wife of Colonel Lake of the Horse Artillery, and a Mrs. Saul, a Carlisle lady, and one or two songs by Miss Hodgson.

Yesterday morning I went to Dalston, about five miles off, to visit mills, and visited some more mills here, and a large school attached to the mill of Messrs. Dixon.   They are the most considerable people here, and very wealthy.   They have lately built what I may, without using an inapplicable term, call a magnificent cotton mill.   The chimney is in the form of an octagonal tapering prism, standing quite detached from the rest of the building, thirty-two feet diameter at the base, and rising to the astonishing height of 305 feet.   When it was finished, but before it was used as a chimney, it was visited by vast numbers of persons.   Mr. Dixon had an apparatus, by which parties were drawn up the inside of the tower, by fours at a time, and when they got to the summit, they found seats and a platform with cake, wine, and fruit.   This was, of course, for Mr. Dixon's friends only—Miss Hodgson went up. There is a very fine prospect from the summit.   Adjoining the mill, the Messrs. Dixon have built, at their sole expense, a capital school, capable of teaching 300 children, open to any one on the payment of twopence per week, and have engaged a very excellent master, and are immediately to have a mistress ; I stayed an hour in it yesterday with Mr. Peter Dixon, and was much pleased with all I saw.   The master is a most sensible man, and is going to introduce geography, mathematics, the elements of astronomy, drawing, and music, and he says that Messrs Dixon second him readily in every-thing.   The pupils will get all these branches for twopence a

week. When I came home I found Mrs. and Miss Hodgson
had called upon me in their drosky, to ask if I could drive out
with them, and left a message for me to breakfast at the
Deanery this morning.   I went there, had a pleasant talk of
an hour.   This forenoon I have been at Messrs. Dixon's
making a minute enquiry about the condition of the hand-
loom weavers.   A commission has been recently issued on
this subject, in consequence of a vote of the House of Commons
last Session, and I promised Samuel Jones Loyd, who with
Senior is one of the Commissioners, to give him some
information on the subject.

God bless my dear Anne.

———

To his Wife.

Manchester, *October* 30*th*, 1837.

I had the happiness, my dearest Anne, of receiving, on
Saturday your dear letter of the preceding day. . . . .

I go on very comfortably here with my factory people.   I
meet with the greatest civility everywhere, and they are
becoming more and more reconciled to the Act, and therefore
I shall be able to do more and more good to the poor children.
I am striving with the difficulty of ascertaining the *real* ages
of the children from physical characters, and have had con-
sultations with many doctors upon the value of the teeth as
a test, I mean the growth of the second teeth, and I believe,
from all they say, that it is the most unerring we can use.   I
am becoming rather knowing in that way for I have looked
into 500 little mouths lately. I suppose it has got wind, for
when the doctors and I go round the mills, and call any to
us that appear too young for their work, they sometimes
come running up with their mouths open, and turn up
their little heads without being told ; I was in a mill this
afternoon, where I found forty-five children collected in a room

at the top of the building, little ragged creatures, but getting some excellent instruction—this they probably would never have received, but for the Factory Act. As I was going upstairs I heard singing, and when I got to the school-room door, it was these little creatures chanting " God save the Queen." When I got into the room I found that they were singing the words of a hymn to that tune. A clergyman of an adjoining church came to examine them while I was there. An hour afterwards I found them busy as bees under the spinning frames. I have a good prospect of getting a large factory school established here ; I have already got five mill-owners to agree to it.

My love to everybody.

My dearest Anne,

Your affectionate husband,

LEONARD HORNER.

CHAPTER XIV.

1838.

To his Daughter.

Bury, Lancashire, 22nd _February_, 1838.

MY DEAREST LEONORA,—On Saturday last I called on Mr.
Heywood* at his Bank, and he took me to Claremont, where I
stayed till Monday morning.   We had a sociable evening and
Mr. Heywood read a part of "Waverley" to the assembled young
party.   We went to the parish church at Eccles next morning.
On Tuesday I went to a meeting of the Philosophical Society,
where I heard an interesting paper by my friend Mr. William
Greg, on the Cyclopean and Pelasgic mural remains in Italy
and Greece, some of which he had examined and made
drawings of on the spot.   Yesterday I went to Patricroft,
beyond Eccles, and besides visiting a factory there, went to
the great iron foundry and machine makers' establishment of
my young friend Nasmyth, son of the celebrated landscape
painter at Edinburgh, an old pupil of the School of Arts, and
a most ingenious and meritorious person.   He is in a fair way
to be the most considerable person in his way in Manchester,
and is respected by all who know him.   His sister keeps house
for him, and I visited her also.   Her room was hung round
with very interesting landscapes in oil, of her own painting,
and over the chimney-piece a most excellent picture by her
brother James, done at moments he could snatch, of an
alchemist in his study after the manner of Gerard Douw.   Mr.
Heywood, who is all kindness to rising merit, or to merit
endeavouring to rise, was good enough to say he should like to

Afterwards Sir Benjamin Heywood, Bart.

know him, and commissioned me to ask him to meet me at his house at dinner next Wednesday, which has gratified him very much, as Mr. Heywood is so beloved and looked up to. He is the President, as he was the founder, of the Mechanics' Institution of Manchester ; having been in bad health for a long time, he had not been much with the directors ; but being better, and having something to settle with them about their anniversary next Wednesday, he invited the whole body, twenty-four, to sup with him at Claremont last night, and he asked me to come and help him to entertain them. He had two omnibuses to bring them up from Manchester. The only other person, besides myself, who was not of their body, was Mr. Samuel Robinson, Mrs. Heywood's brother. We sat down at a long table most handsomely and abundantly provided. We had toasts, and speeches, and songs, which lasted till one o'clock, when they all drove off again in their omnibuses to Manchester, giving me a corner in one of them. They are almost all young tradesmen and clerks. They conducted themselves with great propriety, and some spoke admirably well. The only singer was a professional man, who teaches music in the Institution. I had to make a speech, for Mr. Heywood gave my health and said many flattering things of what they had learned from the School of Arts, and I am enrolled as a member of the Institution. Mrs. Heywood and her children came into the room after supper, and heard the singing and the speeches.

This morning I came here, and have been hard at work all day.

God bless you, my dearest Nora,
Your affectionate father,
LEONARD HORNER.

To his Daughter.

Manchester, 1st *March*, 1838.

My Dearest Joanna,—I have been occupied all the morning in visiting factories and yesterday I dined at Claremont where Mr. Heywood had a large party assembled to accompany him to the general meeting of the Mechanics' Institution of which he is President. We had but a short time for dinner, as we had to be in Manchester before eight. I took Mr. Nasmyth with me. Mr. Heywood read a very beautiful address, quite characteristic of his good sense and benevolence, and there were many speeches, some of them good. Mr. Heywood had a sentence about me, in kindly terms, alluding to some hints for the improvement of their plans which I had given him, and he said that he was glad to say I was then present, and that I had enrolled myself as a member of the Institution. This I have done from my official connection with Manchester, and from my confidence in the good which the Institution is doing. I have got Mr. Nasmyth to be a member, and I hope by-and-bye that he will be made a director, for if so his excellent sense and experience will be of the greatest use.

My dearest Joanna's affectionate father,

Leonard Horner.

———

To his Wife.

Manchester, *March* 11th, 1838.

I have been thinking of late what you have often urged, that I should take in hand myself to draw up some memoir of my brother, from the valuable materials for that purpose which are in existence. I have some thoughts of doing something in this way. To have all the papers and letters copied in chronological order, connected together by such a thread of narrative as may be necessary for explanation. When that is accomplished, it will be a matter of consideration how far the publication will be advisable, and when materials are all brought together in that accessible form, I can the more easily take the opinion of others, as to the desirableness of publication. I shall try to get testimonies from his contemporaries, and I daresay I shall in some instances succeed. Pillans for his boyhood, Jeffrey, Cockburn and Murray for his more advanced years, and Hallam and John Allen for his political life. I shall look for your assistance and of my dear girls as copyists, I am sure that this will be cheerfully undertaken. I have written to Jeffrey to send me forthwith all the papers which I left with him.

During the last week I have been reading the fifth volume of the life of Sir Walter Scott, and it has almost superseded all other reading. It is very interesting, and very curious, and I have not in this volume found anything that I could have wished Lockhart to have omitted, as I did in several instances in the fourth volume. How much excellent sense and excellent feeling is displayed in his letters to his son. There is also less of politics, which is a great relief. How much the kindliness and generosity of his nature were tarnished by his political animosities. Considering the part he took, I must do Lockhart credit for speaking of the Beacon newspaper at all, though indeed he could hardly avoid it; but there is not one expression of regret for the slander and calumnies it poured forth on many good men. Scott, if he had pleased, might have put an entire stop to it at any time; there is not a question that he gave silent countenance to it.

From Mr. Francis Jeffrey.

Edinburgh, 15*th March*, 1838.

MY DEAR HORNER,—We are all coming up to town bodily next week—and as I send a servant with some things by sea, I shall put your box under his care, and bring the key to you myself.

I cannot recur to the subject of that box without melancholy recollections, and while I cannot look back upon my own dilatoriness and inefficiency without some self condemnation, I can most truly say that these infirmities were not abetted on this occasion by an indifference or want of interest on the object in view. Had I cared less, indeed, about the suitable execution of the task, I should have proceeded to it with less hesitation. But I was overpowered, partly no doubt by the mass of materials, but chiefly by the deep sense of the difficulty of satisfying myself, or those I was most anxious to satisfy, in the accomplishment of what I had undertaken. I am rejoiced to find that you think now of putting your own hand to the work.

Any little aid which I can give, I shall be proud and happy to furnish, and though I can only be reckoned on for a slight sketch, I have still so perfect a recollection of the original, as to be pretty sure of its fidelity. I learn with great delight, but with very little surprise, that the memory of my lamented friend is still so honorably cherished, even by those who were strangers to his person; a fame so pure as his, is not liable to be eclipsed by a rapid succession of minor candidates, and during the long period which has now elapsed since his death, I do not recollect an individual who has come upon his ground.

I expect to be in town about the 25th, and shall probably be in the lodgings I had last, 6, Arlington Street. With kindest remembrances from all our little household, to your large one.

Believe me always,
Very faithfully yours,
F. JEFFREY.

———

[The following letter from John (Lord) Murray, the great friend of Francis and Leonard Horner, on the death of his only child, a boy of ten years of age, was sent to Mrs. Horner, and found among her papers after her death. It was probably addressed to one of his constituents at Leith, for which he was then M.P.]

126, George Street, *Wednesday, April 18th*, 1838.

MY DEAR FRIEND,—When I reached Leith on Saturday morning I was in hopes that I might have conversed with you soon on some urgent matters, for before I landed, a report reached me that my son (whom I did not believe to be in danger) was greatly better. When I came to his bedside he was delighted to see me and conversed with ease and playfulness and with intelligence beyond his years, on a variety of subjects. One symptom alone alarmed me, and betrayed the lurking, deep-seated disease which carried him off soon after eight o'clock that evening.

You had many opportunities of observing what he was, kind and affectionate, at times amidst all his mirth and vivacity, prematurely discriminating and reflecting; his disposition to assist boys whom he thought ill-used by others, or who found their lessons difficult, endeared him to his companions, and to all around him. His voice was sweet and powerful in a degree rare at his age, and united to native grace of manner. His desire to excel in all he was taught was so great, that I found it necessary to restrict his attendance at school, and rather restrain than encourage the too rapid progress which his eager energy made, though apparently with little labour to himself. Such was my dear boy, over whose remains I saw the grave closed two hours ago, but my loss is as nothing compared with that of a tender mother, devoted to him from his birth, and who is now bereaved of her only child. The Author of all good gave us much, he has taken away what we loved most, and it would

be a crime to repine at what his unerring wisdom has
ordained.

What remains for me during my pilgrimage here, is to
endeavour to make myself useful as the citizen of a free
country.   To merit the confidence of those whom I represent
(which, I believe, I have hitherto possessed), and to discharge
every public duty, are the only ambitious pursuits which can
now occupy my mind or form motives to exertion.

I am desirous (after having one day's repose in the country),
to see you and any friends who may wish to converse with me
on business, but I entreat that then and for some time, no
allusion may be made to what is hardly ever absent from my
thoughts, but which has inflicted a wound too recent and
severe to be laid open without agonizing pain.   Even the
best selected topics of consolation occasion great sufferings,
and my friends will soon feel assured that no efforts are
wanting on my part.   They will observe that business with-
draws me from sorrowful recollections, and is for the time a
relief and consolation.   In that respect I require no greater
indulgence than they have hitherto shown me at all times.

Although education has of late years been greatly improved
in this country, and is as good at Leith as in any other
seaport, much may be done there, as well as elsewhere, to
improve the morals and promote the welfare and happiness of
early youth.   It is a subject which I have studied for many
years, and as any knowledge or experience which I have
acquired, can no longer be of use to one who was the object of
my care, and nearest to my heart, I will, as soon as I am
released from my attendance on Parliament, direct the
attention of my friends to it, and do what I can to promote
the Cause of Education throughout the district.

I ought, perhaps, to explain why this funeral was more
private than is usual is Scotland.   I suffered much a few
years ago when it was thought right to ask the attendance

of those who for years had known and respected my aged parent. On this last more trying occasion, I had some distrust whether I could lay my dear boy's head in the grave without showing weakness and feelings which might be painful to those present, and if any of my friends from Leith had come, a great part of a population, who justly consider me as one of themselves, would probably have attended. Such funerals are often in this climate productive of serious injury to the living, and approach too nearly to an ostentation of woe to be the best mode of showing regard for the dead.

I remain, my dear friend,

Yours sincerely,

JOHN A. MURRAY.

———

To his Wife.

Delph, *May 20th,* 1838.

I have been thinking of you and your party all day. I hope you have had fine weather to enjoy the last of your pleasant days at Cassio Bridge. I was very glad to learn that Darwin was to go with Charles (Lyell) and Mary to-day. Remember me most kindly to him; tell him that his book has been a very great treat to me. Those only will fully enjoy that book who have the advantage of knowing him personally; I fancy him talking to me all the while. I never read a book in a more conversational style, and it is quite delightful to me on that account. He comes out now and then with some of those exclamations which are so natural in talking, and so seldom written, and there is every now and then a piece of quaint humour which is very amusing; for instance about the number of hen ostriches trying to persuade an old cock to sit upon their common deposit of eggs, and the little beast that foolishly bores holes in a wall, and is marvellously surprised when he comes at the daylight on the other side. What a number of privations he must have endured in travelling through the country between Bahia Blanco and Buenos Ayres, and afterwards to Santa Fé, of which he says nothing. When I meet with a bad, musty and frousy bed, I shall think of Don Carlos and lie down with content. I have got as far as Port St. Julian.

To his Daughter.

Whalley, May 26th, 1838.

MY DEAREST FRANCES,—I went from Manchester to Haslingden on Wednesday, a small country town among hills in the centre of a large mill population. Accrington is a clean pretty village with many handsome stone houses. A Mr. Hargreaves is the great proprietor here, and a very benevolent man; he built a church lately, entirely at his own expense. He has very extensive works for printing calicos, and is erecting a cotton mill on a very large scale. I returned to Haslingden, visiting another large factory, Mr. Turner's of Helmshore, another worthy, good man, who does much for his people. He has 108 children working in his mill, and going to school, upon a plan of relays, which he has had in operation nearly since the Act began, and both he and his people told me that the Act gives them no trouble, that it works quite well, and Mr. Turner said that he was perfectly contented with it. He employs above 1,000 people. In the evening I went to Clitheroe; Whalley and Clitheroe are situated in the upper part of a very wide valley, Ribblesdale, watered not only by the Ribble, but by two other fine streams, the Calder and the Hodder, which fall into it not far from hence. The scenery on the Hodder about ten miles from hence, is, I am told, beautiful. They call it here their Switzerland. I visited four mills in Clitheroe this morning, one of them belonging to a Mr. Garnet, upon a large scale. Near this, about four miles off, is Stoneyhurst, the celebrated Catholic seminary, a place of which I had heard much, and I desired much to see it. I consulted Mr. Garnet and he most kindly said he would accompany me, because, knowing the President, he might be able to procure me a more complete sight of it than I should get by going alone. I called for him at his private house, a pretty spot, where I found him sitting (in a parlour lined with oak panelling saved from the demolition of a part of Whalley Abbey) with a niece, a nice ladylike person, and two sweet little dolls of grandchildren, with flaxen hair down to the middle of their backs, in brilliant ringlets. Mr. Garnet is a very gentlemanlike man. He went with me in my chaise, and ordered his servant to follow with his gig, as we were to part company, he to go back to Clitheroe, and I to proceed to Whalley. There was an English Catholic college at Liège, of the order of Jesuits, founded in 1609, and at the time of the French Revolution, in 1793, they were all forced to fly for their lives, leaving nearly all their valuables behind. They came to England in search of an asylum. Stoneyhurst was an old mansion belonging to Mr. Weld of Lulworth Castle in Dorsetshire, a Catholic, and father of the late Cardinal Weld, which was only inhabited by his steward. He gave it, and a small portion of land, to this Jesuits' college. They had considerable property in the English Funds, with which they bought more land, and built a large addition to the house. It is now the greatest Catholic seminary in England, and has been rising in importance and wealth every year. There are two departments, one for boys, a boarding school; the other for the education of young men who are to enter into holy orders. The sons of the great Catholic families, both of England and Scotland, are sent here; at present there are about 170. Shiel, and Wolf (the present Solicitor-General for Ireland), were educated here. I was very much struck with the front of the building, as we drove up a long walk, with broad grass plots on each side. It is of the style of Inigo Jones, and built

in his time, but he was not the architect. The president was
engaged, but he sent a message that one of the gentlemen
would attend upon us. Very soon a gentleman with a black
gown, and long false sleeves hanging down behind, but with a
hat, came to us, and received us most politely. He was a Mr.
Rowe, as I afterwards learned. He conducted us first to a
long gallery; as we passed through a corridor, I saw several
of the Fellows, in black gowns, and three or four had those
lofty caps which one sees in the pictures of the Jesuits.
There is a large collection of pictures in the gallery, a few of
them very good, particularly some by Domenichino. He then
conducted us to the church, but the sacristar could not be
found, so we went to the gardens, and there we found him in
a scull cap, playing at bowls with some of the lay brothers,
on a beautiful green, surrounded by a noble yew hedge. We
crossed the play-ground, which contains, I should think, two
acres, and in the centre of it there is an isolated lofty stone
wall, about eighty feet wide, for the game of tennis. There is
a handsome observatory in the garden, built two years ago,
and well provided with instruments. The church is a beautiful
building, something in the style of the body of St. John's
chapel at Edinburgh, but with four pinnacles at the angles;
it is in far better proportion too, both inside and out. The
interior is very handsome indeed; there is a fine window
with modern painted glass, which cost eight hundred pounds,
a very grand altar, and a large organ. The whole cost fifteen
thousand pounds, and was built a few years ago. Mr. Rowe
had the goodness to show us a great many things which are
not exhibited to all strangers. We saw a most superb
collection of priestly and other sacred vestments; the word
splendid is not too strong. Among them there is a complete
set, in high preservation, which were given to Westminster
Abbey by Henry the 7th. It came into possession of the
College at Liège, through the Shrewsbury family. When the

Portuguese chapel in South Audley Street, was given up, the
whole wardrobe, which was rich and extensive, was purchased
for Stoneyhurst; it had been presented to the chapel by the
King of Portugal, and was of the most costly description.
The word vestment is applied not only to what the priest
wears, viz., the Cope, the Chasabel, the Stole, and the Maniple,
but also to the veil which covers the cap, and two other flat
pieces like large kettle holders, which are used in the service
of the altar. Some of the richest, and most gorgeous
embroidery, had been restored by a lay brother, who is also
the chief tailor of the establishment, now in the house. We
next went to the library, where we were shown some
beautiful and curious things. A painting of our Saviour on a
wooden cross, by Michael Angelo, about a foot high, and
of the most exquisite finish, also a carving in ivory by Michael
Angelo, of Christ on the Cross, and a stand for the cross with
a group of the Virgin and others. There are the beads of Sir
Thomas More, which are a kind of filagree work, very like the
Berlin iron necklaces; his George, and a gold cross, with four
very large pearls that belonged to him. There was also the
private breviary of Mary Queen of Scots, given to the
Arundel family by her chaplain. We were shown a large
collection of communion plate, and decorations of the altar,
chiefly silver gilt, and some of solid gold; some of them are of
beautiful workmanship. The arrangement for the boys are
admirable; the greatest neatness and comfort, and exquisite
cleanliness. I never saw anything better arranged than the
dormitory, every boy has a little cabin to himself, like those
of the pensioners at Greenwich. Mr. Rowe gave us a very
nice luncheon in his own rooms, and I came away very much
gratified with all I had seen. I must not forget to say, that
the college stands on a rising ground commanding a view of
great extent over Ribblesdale, bounded by the lofty Pendlehill,
and the distant Yorkshire hills. I asked Mr. Rowe about the

works in which the best account is given of the Jesuits, and he
told me that the life of their great founder, Ignatius Loyola,
by Bonhours (Vie de St. Ignace), a work greatly extolled by
Voltaire, and Serruti, *Apologie de L'Institut de la Societé*, are
the best. There is a fine portrait of St. Ignatius by Velasquez
over one of the altars in the church. There is a most,
valuable library bequeathed to them by the late Lord Arundel.

I got here to Whalley about five o'clock, and it was a most
lovely evening. I walked out to view the ruins of Whalley
Abbey, which are extensive, but not very good, and I strolled
for two hours on the lovely green banks, and in the meadows by
the side of the Calder, wishing I had you all with me, not only
for the sake of the lovely green fields, but because I should
have seen you all devouring with the most intense interest,
the many curious things I saw in my visit to Stoneyhurst.
Good night.

To his Daughter.

Ulverston, 10*th June*, 1838.

MY DEAREST JOANNA,—I always admire with increased
pleasure the approach to Lancaster, the features are so grand,
and the bold outline of the Westmoreland and Cumberland
mountains was very marked. I got there between three and
four and went out to visit the mills. Towards sunset I walked
to the Castle to enjoy the fine view. Yesterday, after an early
breakfast I went to Dolphingholm, about seven miles south-
east of Lancaster, in a part of Wyersdale, a pretty wooded
valley.

It is a large establishment for the spinning of worsted,
carried on by Hindes and Derham, and it is one of the best
conducted in my district. The mill is surrounded by the
cottages of the work people, and they form quite a community
of themselves. There is the most delicate cleanliness observed
in the mill and all about it, and the whole group of houses

are fresh whitewashed outside nearly every year, and contrast well with the bright green of the trees. The moral condition of the people is a great object of attention with the proprietors, they pay eighty pounds a year to a clergyman, have built a chapel and school-house, and maintain a schoolmaster. From all I hear, it is a most virtuous and happy little colony, one of the instances where a well-managed factory, under the guidance of enlightened and benevolent proprietors, is a blessing to the country. It would be impossible to collect a set of more healthy, tidy, orderly people, than those I saw working in the mill.

I returned to Lancaster and soon after got on to Burton. The full sea of Morecombe Bay with the distant Furness Fells, were in view all the way. Near Burton I had to visit a large factory situated at the village of Holme, and in a snug, pretty. nook. It is the property of Messrs. Waithman, Quakers, and is for the spinning of flax; another well-ordered establishment, where I found thirty-two children gettting education under the obligation imposed by the Factory Act for that purpose. Mr. Waithman asked me many questions, and took notes, about plans and books for the improvement of his school. A chaise conveyed me to Milnthorpe, about five miles, where the Ulverston coach took me up; I had not been half a mile from Milnthorpe when I was struck with the uncommon beauty of the country, particularly the village of Heversham, where Lord Howard has a fine old place. I stopped the coach and got outside. I had nothing to wish for but a temperature, more like June, and less like March. But the fields and trees had their verdant clothing, in the greatest freshness of new attire. I have seldom had a more beautiful drive. Good-bye, dearest Joanna,

Your affectionate father,

LEONARD HORNER.

[As Lord John Russell announced, in answer to a question from Lord Ashley, that the Factory Bill would not be proceeded with that Session, Mr. Horner felt able to go with his family to France for a few weeks, and visit his sisters, who resided in Paris, and whom he had not seen for more than seven years.]

To Charles Lyell, Esq.

Rue de Ponthien, Paris, *July 24th*, 1838.

MY DEAR CHARLES,—I was at a meeting of the Academy of Sciences yesterday. I spoke to Arago, Brogniart, Adolphe Brogniart, de Beaumont and Cordier. Becquerel was in the chair, and I heard Poisson, Lacroix, Thenard, and Bory St. Vincent speak. Brogniart read a report upon a communication from Colegno respecting the tertiary formations around Turin, and I heard with pleasure in a report read within the walls, references to "les Epoques, Eocene, Miocene et Pliocenes de Lyell."

De Beaumont read a report upon a communication from Bobbaye respecting the geology of the country around Constantine. On Saturday we were at St. Cloud and Versailles, on Saturday night at Franconi's horsemanship and were delighted, the best I ever saw without exception, but I never saw Ducrow. On Sunday evening we walked on the Boulevard des Italiens for a long time. Yesterday we drove in Power's carriage round the Boulevards to the Jardin des Plantes, and after staying there some time I was set down at the Institute. I went home and took Mamma and the girls to dine at the Café Richard in the Palais Royal, with which they were much pleased. We then proceeded to Le Théatre Français, it was "Le Mariage de Figaro," and Mdlle. Mars played Suzanne. It was a great piece of luck to see that celebrated actress, and we were much pleased with her. The play was for the benefit of some poor people who have lost their property by fire.

The more I see of Byrne the better I like him. They are all well, and are unwearied in their kindness. With love to Mary.

Most affectionately yours,
                    LEONARD HORNER.

To Charles Lyell, Esq.

Clermont, *September 4th,* 1838.

MY DEAR CHARLES,—We have completed our visit to Auvergne, and set off for Paris to-morrow morning. We have all to thank you for the recommendation, as we have had a great deal of pleasure, and I have seen a great deal of interesting and instructive geology. We have had a fortnight free to give to the country independently of the journey from Paris, a period far too short for anything like detailed examination, or even to see the great facts, but I have got a knowledge of volcanic phenomena which I had not before. I have seen Gravenoire, Volvic, Nugère, Pariou, Petit Puy de Dome, Pont Giband, Chaluzet, Gergovia, Pont du Château and the Puy de la Vache, with the Lac d'Aidat, and from these I have been able to form a pretty correct notion what craters and coulées of lava are. At Mont Dore we saw the great features of the Valleé des Bains, the Lac Pavin, and the remarkable isolated rocks, La Tuilière, and Sanadoire. I have seen so little, that I am not at all entitled to criticize the opinions which writers in this country have expressed, but I have this remark to make, that it appears to me that all have represented Mont Dore as far more simple in its structure than it is, and have drawn general conclusions from too limited observation. I have not been able to discover any appearance which lends the least support to the theory of elevation craters. The mass of Mont Dore is broken up by rents and ravines which run in various directions, and without a detailed examination of each of these with careful

Y

levelling and determination of points of bearing, I do not see
how a just opinion can be formed as to its mode of formation.
A model upon a pretty large scale would throw a great deal of
light upon it, but it would take an experienced and skilful
geologist many months to make such a model. I have made
a collection of about one hundred and fifty specimens of the
various forms of volcanic products which I have met with. It
appears to me that the eruptions which are said to be of very
different ages present almost an identity of character,
mineralogically speaking. My desire to visit Vesuvius is
greatly increased by what I have seen here, and in order to
understand this region of extinct volcanos, one should have
most narrowly observed volcanic agency in active operation,
I shall indulge the hope of being able to make a run to
Vesuvius one day.

On the morning of the twenty-eighth we made a party to
visit Le Puy de Pariou, &c., accompanied by Lecoq and
Bouillet, who are very kind obliging men, and know every
step of the country. We started in two char-a-bancs soon
after six, and after a merry breakfast at La Fontaine des
Bergers, at the summit of the plateau, where we had sundry
jokes and much laughing about our *trois services* of *Omlette*,
we began the ascent of the Pariou. It was not long before we
were on the edge of the perfect crater, which drew forth from
all a shout of admiration; we went round the edge and saw
the external crater, or at least that which remains, like the
Somma of Vesuvius, we also saw part of the great *chiere* from
Le Puy de Dome, Le Puy de Chopine, and all the wonders that
lay around. We made a quick descent. We had then a
walk of about a mile across the plain which separates Pariou
from the little Puy de Dome, and there we sat down on the
edge of that crater, *le Nid de la Poule*, which is nearly as
perfect as Pariou. We had intended to have gone up to the
summit of the Puy de Dome, but the day was too far advanced

and it was very hot. Brown, however, went, and we sat down on the slope, picking bilberries, which we found in great profusion. Brown came back sooner than we expected, a gentleman *Chasseur* whom we found on the mountain, a friend of Bouillet, acted as his guide. The char-a-bancs were waiting for us at a tree near the foot of the Puy de Dome, and we had a thorough shaking in crossing the lava stream of Pariou to get into the cross road. We found some capital bottled beer at La Baraqne, and got home about six, much pleased with our day's excursion. Next morning we started for Mont Dore, we hired a carriage, a sort of double post chaise, to take us all the way, it is about thirty miles from Clermont. In four hours, at eleven o'clock, we reached Randanne, the house of Count Montlosier, for whom I had a letter from his very old friend Mallet. I have omitted to say that all the time we have been here, and in all our excursions latterly, we have had the pleasure of the society of Mr. Melville, who brought me a letter from Mrs. Somerville. Count Montlosier received us with the greatest kindness. He spoke of your visit to him with the Murchisons, when he received you in a cottage, but now he has a capital house, well furnished, but smelling strongly of *Cowigen*, as Dr. Beddoe's patient called the air of his byre that he made her breathe; for the Count has the wings of his house occupied by about thirty cows in each, with a free communication with the centre. He had a capital breakfast ready for us, and after spending about three hours, we left him with a promise to see him again, and spend a night on our return. We did not reach Mont Dore les Bains till eight o'clock, so we had the disadvantage by the darkness, of losing the scenery of the approach to the valley. The people of the inn had been prepared by Lecoq for our arrival, and we found capital beds, and a capital dinner. Next morning, the 30th, we started at nine to visit la Tuiliere, and la Sanadoire. We were all

mounted on ponies with the exception of Brown, who preferred
walking, and the ladies had excellent side saddles. We
arrived, after crossing several hills and wooded valleys, at the
object of our expedition, about one o'clock. I do not know if
you visited this spot. They are two insulated rocks at the
entrance of a deep combe, rising to the height of about 500
feet from their base, composed of *Clinkstein*. La Tuilière
presenting a most beautiful group of columns, which put me
in mind of the Island of Eigg. We stayed above an hour on
the spot, and returned the way we came, much delighted.
Next morning, the 31st, we started at seven, ascended the Pic
de Sancy, and went as far as the Lac Pavin, a beautiful crater
lake, deep sapphire water, with the sides clothed with trees.
Here we took our repast, which we had brought with us, and
returned the way we came, reaching our inn at seven. When
you consider that we were ten hours on horseback, riding up
and down steep ascents, over very rough stony places, you
will not be surprised that we were a good deal fagged, Mamma
and I most, Frances the least of the party. It was a very
hard day's work, but the scenery in many places sublime.
Neither Mamma nor the girls shewed the least fear, passing
on horseback along very narrow paths on the steep sides of
the mountain, so much confidence do these nice Auvergne
ponies inspire. On the 1st, Mamma and Susan stayed in the
village. Frances, Brown, Meville and I went to the Gorge des
Enfers, and the valley de la Cour, where we were much
interested by the grand display of dykes, traversing the
conglomerates, &c. We saw the establishment of the baths by
the chief physician and director, Dr. Bertrand, for whom I had
a letter from Arago. We started at six in the morning of the
2nd, and at ten were again at Randanne. After breakfast
the Count got out his two char-a-bancs, and we set off for the
Lac d Aidat, where we had the opportunity of seeing to great
perfection the vast stream of lava, flowing from the Puy de la

Vache, which choked up the streamlet to form the lake, and flowed on for three leagues. We got back to dinner at five, and spent a very pleasant evening with the Count, who amused us with many interesting stories connected with the early part of the Revolution, when he was flying with the emigrants after the defeat of the Duke of Brunswick. He is a very fine old gentleman, with most graceful manners—a fine specimen of the old French noblesse of the best sort. He is eighty-three, reads without spectacles small print, and not only has every one of his thirty-two teeth, but strange to say, within the last three months, three more have sprung out at the side of his jaw, and when we were there, he told us he was suffering considerable pain from the cutting of a thirty-sixth tooth.

Yesterday morning Brown, Meville and I rose at six, and visited the Puy de la Vache; after looking at the outburst of the lava stream in the bottom, and the pouring out from the broken down side, we scrambled up the face of the crater inside, no easy work, as the cinders gave way under our feet and we sank ankle deep into them. This is the most complete in its character and resemblance to an existing volcano, that I have seen. We took leave of the Count with great regret after breakfast, and reached Clermont about two o'clock. Yesterday was the first meeting of the Scientific Congress; Bouillet and Lecoq would have us enroll ourselves as members, and this morning at seven, we attended the first meeting of the section of Natural History, when they chose Brown as President of that section, the Abbé Croizet and Professor Maravigna of Catania, as Vice-Presidents. A botanical paper was read by a Doctor Bravoz, which Brown says is a very curious one, and may excite some attention. The subject of it is to show that the leaves and component parts of the flowers of all plants are arranged according to mathematical laws. When we were at Paris, the savants of

the Institute either did not know, or affected not to know, of
the existence of this scientific association, not one of them is
here, and Becquerel, who was here last week, went away.
So much for the jealousy of the Parisians of anything like
distinction existing in the provinces. Except Caumont of
Caen, who is President, and Bouillet and Lecoq, there is not a
person whose name I ever heard before. They are equally
unknown to Brown. Frances has had the greatest enjoyment
in her botanical pursuits, she has been unwearied in her
exertions and has made an immense collection. When we
were all sleepy and tired with our day's work, she would not
go to bed till she had fixed all her specimens in sheets of
paper. You may judge from this that she is remarkably well.
She has had great kindness and assistance from Brown and
also Lecoq.

Susan has been no less active with her pencil, and was
generally up as late as Frances, finishing with sepia her
pencil drawings. Our places are taken in the diligence for
Boulogne on the 10th, and if no interruption occurs, we shall
be in Bedford Place on the 12th. Nothing must detain me
beyond the morning of the 17th, and as I hear the railroad
to Birmingham is to be opened all the way on that day, I
hope to sleep in Manchester on the 17th. It will be a great
comfort to me to see my three girls before I go, as I do not
expect to return till the beginning of December.

Yours most affectionately,

LEONARD HORNER.

———

To his Wife.

Manchester, 18*th September*, 1838.

I had not proceeded a hundred yards from our door when
the spring of the coach broke, and the man had to go and
fetch another coach, while I stood by to watch my luggage.

The consequence of this was, that I got to the station just as the directors' train was starting, however, I saved my distance, and Mr. Thomas Tooke spying me, hailed me to a carriage in which he was sitting, and I found in it also Mr. Stephenson, the engineer in chief, whom I was glad to become acquainted with, as the planner and executor of this great work. We started at twenty minutes past seven, and at five minutes past twelve we were at Birmingham, but having stopped at Tring, Wolverton, and Rugby, full three-quarters of an hour in all, we performed the distance of one hundred and twelve miles in four hours, equal to twenty-six miles per hour, and without accident or irregularity of any sort. Had a train been ready to bring me, I should have reached Manchester by five o'clock, but I had to wait till three, and we got here by half-past seven. It is something to remember to have been one of the first to have travelled on the day of the opening of such a work, and to have gone from London to Manchester in twelve hours, waiting nearly three at Birmingham.

———

To his Daughter.

Manchester, *October,* 1838.

MY DEAREST SUSAN,—I have seen no one till yesterday, when I dined at Mr. Wood's at Singleton Lodge, three miles from the town. He is in London, but his son asked me to a quiet dinner with his mother, and he only exceeded the promised number by one, Mr. Tayler, a dissenting clergyman, a very sensible, well-informed man. He lived four months at Bonn in 1835, in the house of Professor Brandis. He came away with the same feeling of respect and regard for the Germans as we did, and loves to speak of them and of their country as we do. There is no reading from which I derive more pleasure than their authors. I read, last night, the fourth and fifth acts of Jungfrau von Orleans. It is a great production, and puts me in mind in many places of the vigour and the imagination of Shakspere. I have taken up as my next German companion, the pure, the amiable, the enlightened Moses Mendelssohn; his "Phädon," a book which may be read over and over again with pleasure and profit, for it is impossible to lay it down without calmer and more elevated feelings.

You are very lucky in having so able an Italian master as Mr. Prandi, and you should study hard while the lessons are going on, and take every opportunity of getting your difficulties explained. Up to a certain point the language is easy, but I have always heard it said, that to understand its niceties and higher merits, is not so light an achievement. It will be an additional source to you all of the most refined independent enjoyment. I am sorry that my acquaintance with it is so limited, perhaps I shall again buckle to, and make myself better able to read their great authors.

END OF VOL. I.

Women s Printing Society, Litd., Great College Street, Westminster.

Printed in the United States
By Bookmasters